Gerhard Friedrich
Viola de Galgóczy
Barbara Schindelhauer

Let's visit Number Land

playfully discover the world of numbers

TABLE OF CONTENTS

Preface

The world of mathematics is so important, so omnipresent in our world and so full of aesthetics - yet far too many of us have uncomfortable memories on their math career in school. This is a pity but could have been avoided to a large extent with a better start.

We have committed ourselves to provide all young children with that good start and would like to introduce you to our early math concept "Let's visit Number Land".Our concept has found wide acceptance in Germany since we initiated it in 2003. Throughout the years we have been enjoying the enduring benefits many children, also our own, got out of this playful, natural and highly motivating concept. It has also been great to see how much Number Land helps teachers in their work. Two scientific studies confirmed the positive effects.

We are convinced that our concept will also work well for you and we look very much forward to your experiences. We would very much appreciate your feedback so feel free to contact us, also in case you have any questions.

Have a good trip!

Yours sincerely,

Gerhard Friedrich
Viola de Galgóczy
Barbara Schindelhauer

Welcome to Number Land

"Imagine, Mommy: Today Number 1 visited us and invited us to travel to Number Land where it lives together with its friends Number 2 to Number 10. We get there by walking over Number Lane that goes from 0 to 10 or even up to 20! In Number Town, each number has its own geometrically shaped garden. We can furnish these gardens with houses, house numbers, towers, and other things that go with each number. We can even bring in our own ideas! Ever and again Hodgeypodgey, a cheeky goblin, messes around, but a nice fairy helps us to sort things out again. There will also be tales and music, lots of playing, active games, creative activities and other things we love! I will even have my own Book of Numbers... Tomorrow Number 1 will pick us up for the first visit to its garden. I am so excited!"

Since 2003, children aged 3 to 6 have been telling about their joyful and inspiring visits to Number Land. The level of fun and great success is expressed in the feed-back from children, teachers, and parents as well as in two scientific studies. This is not really astonishing since "Let's visit Number Land" has been consistently developed out of the perspective of young children: It considers how small children see the world, builds on what they know already, what they can understand, like, and need.

The concept appropriately addresses the evolving interest of small children in numbers and in geometric shapes. It supports children to playfully master the transition from their natural understanding of math to the formal mathematical language used in daily life, school and beyond. Scientific findings from brain research, developmental psychology, educational science, and the didactics of mathematics have been carefully woven into an open concept allowing utmost flexibility but also giving a frame where all individual and different impressions and experiences find their place.

Our goal is to help the children to acquire a profound, structured understanding of the number range 0 to 10 and the numerals up to 20. Studies identified insufficient know-ledge of that as the major cause for frustrating math problems. Yet the children bene-fit far beyond the matter itself: Language, self-esteem, concentration and cognition, motivation and creativity, social, motor and music skills are enhanced as well.

The edge of "Let's visit Number Land" lies within its embracing approach: The concept allows a child to experience the *entire* aspects of numbers with the entire self, i.e. body, mind, and spirit. This way, children develop a multi-faceted, positive image and deep, lasting understanding on which they can build. In a scientific study we could prove that within only ten weeks children aged 3 to 6 gained the mathematical competence and understanding they normally would have achieved only within one year. The same was true for the language competence of the children - a highly important finding since language is the key to education. The findings were also true for underprivileged children with partly very poor understanding of the German language. A second study by the University of Heidelberg supported our findings.

School teachers approve the concept as an optimal preparation for school. In Ludwigshafen, which has many underprivileged children, teachers diagnose a higher base level and more self-esteem, leading to more motivation and participation. Teachers also apply "Number Land" themselves.

Nursery and kindergarten teachers are enthusiastic about the all-embracing advancement of the children. They appreciate how well this open concept can be combined with their curricula and how individually it can be realized. Children usually get completely absorbed in a "Number Land" project and transfer it into their daily play.

Using this book

This book is designed for teachers of children three through six years old. It is a practical guide to planning and carrying out "Let's visit Number Land" in nursery schools, preschools, kindergarten, day care, special schools, and primary schools.

The concept is not a training program but a motivating and structuring frame for the teachers and for the children alike. This allows a high flexibility to meet the specific needs of the target group. On the other hand this requires an understanding of the thoughts and the principles behind it.

The goal is therefore to provide you with everything you need for an individual journey to Number Land and to meet your specific needs. It is up to your objectives and up to the children's constitution how deep and how far you may go and where you put the emphasis. The younger the children the less complex the project will be. The older the children are the more you may want to lower the magic aspect of Number Land.

We suggest you read through the book once to familiarize yourself with "Let's visit Number Land" and then plan your own project.

Part I

Part I elaborates on the scientific thoughts behind the concept: Why the early years are crucial for life long learning, what prior knowledge children should have acquired regarding math before entering school, how we have to go about it to successfully impart the knowledge and finally what a number comprises. Then the core elements of Number Land are described, how they are used and what they communicate to the children.

Part II

Part II gives you concrete ideas for how to go about a Number Land project: Decisions to be made in advance and practical tips on the material help to prepare a project. Since getting started is the hardest part, we give ideas for that but have also other ideas for your trips to Number Land. They are by far not exhaustive but give an idea and may be complemented with your own pool of proven ideas and activities.

Part III

This section contains tales and songs for each number as well as suggestions for games and other activities around the number. Again they are suggestions and you may want to modify or supplement them.

There is no one best way to carry out Number Land but there are important principles that need to be followed. Apart from that feel free to put together a project according to your needs. Get started but keep monitoring how the children react to your schedule. It may be necessary to adapt during the project: Do you have to slow down? Should you shorten the range of activities but rather explore less in more detail? Stay flexible here and grab opportunities that come up.

We recommend preparing a Number Land folder in which you may collect schedules, originals and ideas that worked particularly well. Over the time this will be a helpful source for planning other Number Land projects.

It has also proved of value to step back and to let the children work things out rather than too quickly give explanations. If you have any questions please feel free to contact us.

Part I The concept and the scientific thoughts behind it

Why early math is so important

Mathematical skills are essential for school, really every job, and for every day life. We also know today, that the early years are crucial to develop a conceptual and profound understanding of basic math. Such an utter understanding of the very basic ideas is the basis for all mathematical skills later to come. All boils down to the understanding of the basic numbers 0 to 10 in all their different meanings and aspects.

When entering school, math is the second central subject besides reading and writing. Pain and pleasure directly correlate to how well a child perceives its capability to meet the requirements - thus influencing self-esteem and further motivation to learn.

Children differ strongly when entering school, some easily manage the numbers well above 10, whereas others might still struggle with even recognizing simultaneously for example 4 bricks as a set or do not know that 4 can be divided into 3 and 1.

Already very young children have a basic natural understanding of numbers. A toddler who is offered 1 biscuit on the left and 3 on the right will always turn to the 3 biscuits. Usually around the age of four, children start to refine this natural understanding and show strong interest in exploring the numbers: Suddenly it becomes interesting how old they are, how many days are left until grandma's visit and so on. (At this time, children will already have made many more mathematical experiences, but we will focus on the numbers as one big part of the mathematical understanding). Children then enjoy very much exploring and playing around with numbers.

This curiosity is a strong indication that a child is now ready for the topic and that it is up to us to provide suitable and stimulating opportunities to explore and experience math in a way that suits the children's understanding and view onto the world. So rather than imposing math on children, we are on the contrary obliged to provide surroundings that help children to satisfy their urge to learn playfully.

We must also be aware that nowadays a growing number of children do not show the natural curiosity but that it is blocked by external settings. Nonetheless, the potential is there so it is our chance to open these children, to close the gap, to provide a good basis and to prevent math problems.

Important prior knowledge

A well-received long-term study carried out from 1999 to 2003 by German psychologist Kristin Krajewski revealed five skills concerning numbers a child should have acquired when entering school. *(Publication list : Kristin Krajewski, Vorhersagbarkeit von Rechenschwäche in der Grundschule, Universität Würzburg 2003)* These five factors significantly determine whether or not a child manages the various math topics throughout the first four years of school. The study is therefore highly interesting and helps us to define objectives for preschool math. The factors are:

I. Seriation: Being able to assign an element to its correct place in a row. Example: Placing a straw according to its length; continuing a pattern of pearls or shapes.

II. Comparing sets: A set of four remains four no matter how it is arranged or what it consists of. Example: 4 apples = 4 bricks. 4 stacked bricks = 4 bricks in line.

III. Numbers: Identify numerals 1 to 10. Knowing how a number between 11 and 20 looks like when only hearing the word. Example: "Thir-teen" sounds like "31" but is written "13".

IV. Counting: Being able to count (not recite!) forward and backward, knowing predecessor and successor. Example: The predecessor of 5 is 4, the successor is 6.

V. Calculating: Being able to carry out very simple calculations with concrete material. Example: 2 bricks and 3 bricks make 5 bricks.

Children equipped with these skills before entering school have good chances to avoid math problems. We must, however, keep in mind that other fundamental skills determine the ability to achieve these five skills:
General preconditions are motor skills, tactile-kinesthetic-vestibular cognition, visual and auditory cognition, language skills, spatial and time orientation.
Going with that are experiences with the exploration of material, spatial relationships, classification, patterning, one-to-one correspondence, ordering, shapes, fractions, measurement as well as time and money.
We see that many things predetermine and contribute to the fundamental understanding of math and are therefore helpful and important. The five factors named in the beginning indicate a good objective of how far to go with preschool math.

What science tells us and how we apply it to "Let's visit Number Land"

In this chapter, we summarize the theory behind our concept and how we applied it.
"Let's visit Number Land!" combines findings from four important areas: Brain research, developmental psychology and educational science tell us how we have to go about the matter. Didactics of mathematics indicate the contents, i.e. what we have to transport.

Brain research

Each human is born with virtually the entire number of neurons. It is not the neurons but the connections between them that determine our ways of thinking, behaving, feeling etc. Building these connections is a lifelong process. Yet within the first two years a great number of connections between the neurons are built, many more than needed later on. Then a process of thinning out begins: Those connections needed, consistently remain and are being reinforced whereas others degrade.

This is the neurobiological explanation why repeating, exercising, deepening, differentiating, and generalizing play such a big role in education.

The saying "You can't teach an old dog new tricks" is not really true. One would have to add "... that fast, well, or elegantly". An adult for example who has never played an instrument before will still be able to learn it. But even with hard exercising he will only achieve certain virtuosity. Yet a musician mastering one or several instruments will learn an additional instrument much easier than someone who does not. The reason is that the musician can build on already established networks in his brain.

New knowledge is always being integrated into existing neuronal networks. Quality and level of the new things to learn depend very much on the existent. So existing and prior knowledge determine the learning of something new.

Learning is a self-stabilizing and reinforcing process: The more precisely one knows about something, the easier it is to add related contents. This is the reason for the didactic method of starting with easy, concrete experiences before moving on to more difficult, abstract things. For instance, when having to learn complex theorems, that new knowledge is being connected with the prior knowledge.

So prior knowledge is the most important factor in learning new contents. In the beginning again and again external stimuli are important for the brain development. But then the brain becomes a more and more self-steering system, asking questions, looking for rules and coherences.

Consider that this includes emotional associations! Having made really good experiences with a matter we will be open to learn more and be successful. Our central theme, math, is a good example: Once a child has internalised that math is something scary because the people around say so or tell about their bad experiences, barriers will build up and block learning.

"Let's visit Number Land" consequently takes a child's perspective. It allows the children to build on their prior knowledge, their experiences, their abilities, and on how they see their world. The children can approach the topic in many various ways they like and need. This leads to a dense network of fundamental mathematical experiences full of positive emotions and rich inner pictures associated with math. This close meshed neuronal net builds the basis for all further math topics to come.

Activity supports cognitive learning

We all learn much easier when we move while learning something rather than just sitting at a desk while for example counting out loud. Activity supports our cognitive learning and meets the desire of children to move. Interestingly, findings from neurobiology tell us that we subconsciously feel that something (in this case numbers) becomes bigger when moving forward, and vice versa, when moving backwards, something appears smaller. This sounds like a good explanation for the fact that children who are unable to walk backwards, often have difficulties with subtraction.

This is why a lot of activity is deliberately included in Number Land: Whenever moving to and fro on Number´s Row, building the gardens, playing related activity games, dancing to number songs, lying down to build shapes and so on, the children experience math with their entire body.

Memory: The role of incidences, places, regularity

Real incidences burn into our memory – good ones as much as bad ones will stay on our mind forever and we will even remember details. An incident is something new, exciting, sticking out. We also remember the place where a specific event took place very well. Experiencing something on a regular basis (certain day/time) again helps to memorize because then the attention is focused and people expect to experience or learn something about the specific matter.

Applying these findings on Number Land we make the numbers a real happening for the children: The abstract symbols are living, tangible characters whose home, Number Town is right on the floor in front of the children. Number Land is full of songs, tales, things to do and each number has its own adventure and activities! The trips to Number Land are scheduled so the children know approximately what to expect. This draws the attention to the topic and opens the children for things to come.

Developmental psychology

Until a certain age children regard their environment as living and as having either good or bad intentions: A child does not just hurt itself when accidentally running into a table. It was the bad table that deliberately hurt the child! The feelings are not refined yet but reduced to either completely good or totally bad. In addition, anything magic strongly appeals to the children, they love tales and fairy-tale figures who are again either good or bad.
It is very natural for the children to travel to a country where numbers live in their gardens and houses. Amazingly even for 6 year-old tough boys number puppets are usually still vivid and lovable. The magic about Number Land appeals to the children and so do (in particular) mischievous Hodgeypodgey and the friendly Fairy of Numbers, Forgetmenot.
Developmental psychology requires from elementary education to address certain areas in order to ensure a good child development: Language skills, concentration, cognition, motor skills, social competence, emotionality, retentiveness. In addition it is demanded to offer holistic experiences, i.e. that the child can explore a topic with its entire body, mind and soul. "Let's visit Number Land" meets all these requirements,so it is an embracing concept that goes well beyond math understanding and skills.

Didactics of mathematics

The fundamental mathematical understanding of a child includes many fields: attributes and properties of material, spatial relationships, classification, patterning, ordering, shapes, fractions, measurement (incl. time and money) – and numeration as a big part. “Let’s visit Number Land” focuses on numeration but includes other aspects as well respectively paves the path to their exploration. We concentrate on numeration here.

Numbers are used for many things and the children need time to explore and to internalize them all:

* Cardinal aspect: Numerals are used to represent a quantity of things. It is irrelevant whether we have 4 apples, 4 houses or 4 buttons and how they are arranged (invariancy). Neither does it matter which element is counted first, second and so on, they altogether build one set of 4 elements. The one-to-one correspondence means that each element is counted only once.

It is very useful to develop the ability to grasp sets at a glance, i.e. simultaneously. Interestingly, this is only manageable up to the amount of 5. Any sets with more elements than 5 we have to split up into smaller sets that we then add up. This is why we speak of „the power of 5“ and tend to build packs of five. The reason might be that we have five fingers on each hand.

One way of organising a set is depicting numbers with pips as on dice. Using dice in (board) games facilitates simultaneous capturing of quantities. Yet the children (have to) learn that also dots can symbolize a set.

* Ordinal aspect: Numbers have a definite order when counting 1, 2, 3, 4 and so on. Numbers are also used to express a rank: the first, second, third and so on.

* Decomposability: Numbers can be decomposed into other numbers, e.g. 5 can be broken down into 3+2 or 1+4 or 1+1+1+1+1. It is crucial for children to understand and to be able to apply this. Only then will they later be able to calculate beyond ten and to deal with big numbers. Many math problems go back to a lacking understanding of this!

* Arithmetic aspect: A number gives the result of a calculation such as 3+4=7 or 10-4=6.

* Numbers describe how often something is done: I can jump 3 times, clap my hands 5 times, have to sleep 2 more times until my birthday and so on.

* Numbers are used for measuring: 3 inches, 5 minutes, 80 ounces, 20 cents, 4 days, 5 miles...

* Numbers play a role in tales, rituals and have symbolic or mythic meanings: Think of the magic three or number 13 representing bad luck in many cultures yet good luck in others.

* Geometric aspect: Numbers identify geometric patterns such as triangle, rectangle, pentagon and so on. Even the circle relates to 1 and the ellipse to 2!

* Numbers encode things such as a house, a town, a telephone extension, a player of a basketball team and so on.

Considering all these aspects of numbers it becomes clear how much the children have to familiarize with. Children clearly need time and many different, playful experiences to grow a thorough understanding.

We would also like to point out that numbers have a great psychological value: The numbers are very reliable for the children and thus give them good orientation. 4 things will always be 4! A rectangle will always have 4 corners! 4+1 will be 5 forever!

Number Land offers playful opportunities to experience all aspects in a nutshell. For example Five’s garden lies between that of Four and Six (ordinal aspect). The garden itself is a regular pentagon (geometric aspect) and can be decorated at each corner (one-to-one correspondence). In the garden there is a house with five windows (cardinal number) as well as a Number Tower helping to show the computational ability of numbers and how to calculate. Further more we look for numbers in musical structures and especially in the particularly environment of the children.

Scientific findings and feedback on "Let's visit Number Land"

We have been spreading „Let's visit Number Land" within Germany since 2003 through publications, talks and seminars, with overwhelming positive experiences and feedback from the field. Teachers in nursery schools, kindergartens, preschools, and in primary schools value the flexibility, the contents, the stimulative nature and the supporting effect on the development of important soft skills.

"Let's visit Number Land" has been scientifically evaluated twice with "kindergarten German type" children (In Germany the children attend kindergarten from the age of 3, sometimes even younger, until they enter school. The children go on five days a week, either only for the morning or for the entire day). From 2003 to 2005 Gerhard Friedrich and school psychologist Horst Munz carried out two trials with 3 to 5+ year olds, funded by the Ministry of education, cultural affairs, and sports of the German state of Baden-Württemberg. From 2005 to 2009 Sabina Pauen, professor for developmental psychology at the University of Heidelberg conducted another study with four to five year olds. Both studies involved many children with to some extent severe language problems or from underprivileged families.

Description of the study by Friedrich/Munz

During 10 weeks 46 children aged 3 to 5+ traveled to Number Land once a week. Each week they visited one number for approximately 50 to 60 minutes. Whenever suitable, the respective number of the week was picked up during the daily routine (circle time, creative play, sports, meals etc.). The children had access to the material for free play. In parallel, a control group of 46 comparable children carried on with their normal routine and had no access to Number Land. In a second run, this setting was repeated with underprivileged children, many of them having severe language problems and lacking other skills. The charts depict the results of that second run.

Project and control groups were tested before and after the 10 weeks in which the project took place. The tests comprised tasks such as conceiving and building sets of numbers, colors and shapes, remembering numbers, cognition of details, general reasoning powers. Regarding active and passive language skills, the children made a story out of some pictures, were asked what they remembered from a story that had been read out to them, and they had to carry out tasks of differing complexity.

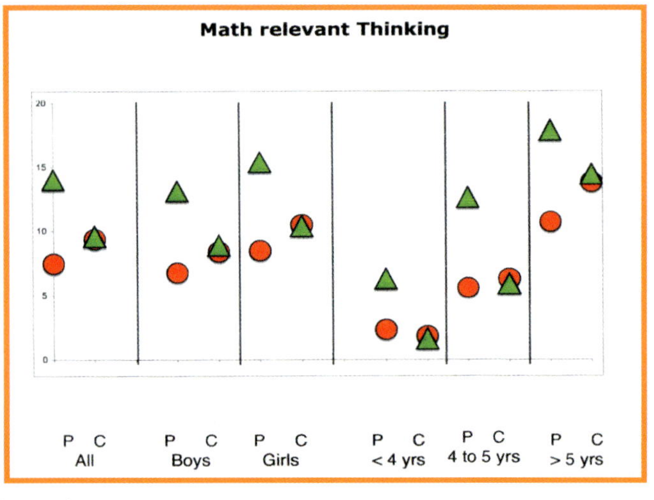

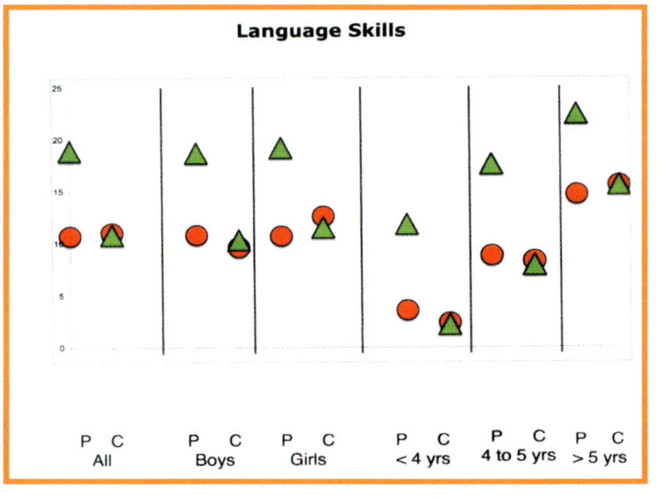

Legend:
P Children participating in "Let's visit Number Land"
C Control group
Red dot Score achieved before the test phase
Green dot Score achieved after the test phase

Findings:

* Only the children participating in "Let's visit Number Land" scored significantly better after the 10 weeks, the control group more or less remained the same.

* Both sexes benefit, the girls even a little bit more than the boys.

* The age specific results show a striking development of the children regarding both math and language skills: Naturally the younger children had a much lower base level than the older (red dots). Yet the tasks were all taken from tests evaluating whether or not a child aged 5+ may enter school. So obviously the older children achieved much higher scores (green triangles).
But after only 10 weeks of Number Land the under 4 year old children scored on average higher than the base level of the one year older! The same was true for the development of the 4 to 5 year olds compared to the base level of the over 5 year olds who again made significant progress.
Summarizing we can say that "Let's visit Number Land" strongly fuels the development of children in two core subjects, i.e. language and math. It is particularly satisfying to see that underprivileged children benefit at least to the same extent. This proves that the gap in education can be closed considerably. Soft skills such as concentration, motor skills, social competence, creativity, and so on were not explicitly included in the studies but the teachers reported progress here as well.

The elements of "Let's visit Number Land"

In this section we describe what the elements of "Let's visit Number Land" are, how they can be used, and what the children get out of them. The elements described here address all areas children love and need, allowing approaches to the subject from different angles. The elements can be combined into an elaborated project but it is also possible to pick those appearing the most suitable for your respective the children's specific needs or interests. It will depend on the children (age, constitution) and on you and your objectives what you will use, how you arrange the elements, and what exactly you will do in the course of the project.

Number Lane

"Number Lane", leading the way to Number Town, is a catchy and appealing description of the number line. It consists of big, robust mats, carpets or cards showing the numerals 0 to 10 (later on to 20) and can be used for many exercises and games.

We include 0 right from the start since thus the children grow easily into the concept of zero. Ideally, you lay out Number Lane directly from 0 to 10. The advantage is that over repetition the children will soon familiarize with all numbers. However, you might as well decide that for your specific group 0 to 5 is the more appropriate for a start. This may make sense when you travel with 3 to 5 years olds to Number Land and only visit the numbers 1 to 5 in a first project. But we see many children of that age who manage Number Lane 0 to 10 very well. You could also offer part of Number Lane first and after a while discuss together with the children whether or not to extend Number Lane. The little ones (up to 3 years) may stick to numbers 0 to 3.

It is always desired that the children have made playful experiences with the entire Number Lane, i.e. from 0 to 20 before they enter school. School teachers report that many children struggle with for example spelling "13" because they hear "thir-teen", which implies that this number is being spelled as "31". This corresponds to the study on important prior knowledge indicating that knowing the symbol of a number behind its name helps substantially when later in school calculating beyond 10. A great game for that topic is "Treasure of Gold".

By laying out the mats in the correct order, moving and playing numerous fun games on Number Lane, children can make many important basic experiences: They become familiar with the numerals, their order, explore predecessor and successor, experience addition and subtraction. They physically experience the numbers and – in addition – train their motor skills.

The most important exercise is to walk forward and backward, ideally counting out lout. The children may stop on one plate and look around before moving on. The important knowledge of successor and predecessor is easily adapted. For fun, the children may also hop on one leg, on both legs or with hands and feet at the same time. Moving forward some steps and stop again, the children make first experiences in addition and vice versa with subtraction when moving backward.

Number Lane can also be used to combine ordinal and cardinal aspect: The children "decorate" Number Lane by counting washers, pennies, pebbles or something else onto each plate. It is a nice ritual to always decorate Number Lane only up to the topical number. Since nothing is put on 0, the children understand that this number is quite useful if they want to express " no counters" or just "nothing". The children may also place the counters in a line next to Number Lane and get the idea that it always becomes one more.

Please see Part II for specific ideas for activities on Number Lane. The ideas can be applied to all type of Number Lanes (0 to 3, 0 to 5, 0 to 10, 0 to 20).

The Number Puppets

The number puppets animate the abstract numbers and appeal to the magic and emotional thinking of children. The puppets communicate that each number is unique, nice, lovable, tangible – and someone to have fun with. They very much contribute to a positive attitude to math.
They look like the real numbers but have features such as a face on them that make them appealing and interesting to children. It is nice if each number has something specific on it, for instance Two has a hat with two feathers, Four has four braids, Five a vest with five buttons on it, Nine a big grin with 9 teeth, Ten wears fingered gloves…
The number puppets live in Number Land and have their own houses and gardens about which they can talk to the children. They may tell a tale, help arranging their garden, talk to the children about themselves, be part of role-playing – and are just available to be hugged and played with in free-play.

Number Town

Number Town is where the numbers live in their characteristic, geometrically shaped gardens with their houses, towers and (always changing) decoration. Starting with a visit to number 1, the children build that garden and explore this number at their own pace. The respective number puppet may accompany the children, introduce, ask and explain things.

Having explored number 1 as detailed as desired, we move on and add number 2 in a following visit, then number 3 and so on. In the end, all ten gardens will be arranged in a full circle on the floor (or on a table). Number Lane runs next to Number Town or leads directly to it. The children can move within this circle and have all aspects of a number range 1 to 10 around them:

The "neighborhoods" communicate the order of the numbers as well as predecessor and successor. Each garden visualizes and interlinks the characteristics of the specific number in different ways. The houses with their windows and house numbers, the towers, the puppets are constant elements but the children can see how these elements change consistently. Finally, each garden and what the children have allocated to it gives a complete picture of the specific number.

A frequent question is: "Where does Number 0 live? Isn't it sad not to have a garden?" The children usually understand quickly that "nothing" has "no" garden. An established explanation is that Number 0 is the very beginning of Number Land, as can be seen on Number Lane with 0 as starting point. Zero is the gate to Number Land. And it is so proud of this important role that it is not at all unhappy not to have a garden on its own…

Geometrically shaped gardens

The gardens correspond to the geometric aspect: Number One lives in a circle that has one centre point. Two's garden is a beautiful ellipse because the ellipse is constructed around two points. As of Three's garden, things become obvious: A triangle for Three, a square (or quadrangle) for Four, pentagon, hexagon, septagon, octagon, nonagon and decagon for Five to Ten.
Children love shapes and by using the corresponding shapes for the numbers we help pointing out coherences. Children enjoy that a lot and will suddenly see shapes everywhere in their surrounding. It is big fun to play with geometric shapes.

Number houses

Obviously, each number needs a home! There are many ways to construct the houses but the principles are important: The windows are depicted like the pips on die. As of Number Six, we have twin houses: One house with 5 windows plus one house with 1 window, then 5+2, 5+3, 5+4 and 5+5 windows for the other numbers. This is easy to understand and quite close to the children since their hands have 5 fingers each. The children may place the twin houses next to each other or on top of each other.

The design of the houses is of very high value: The windows help to understand numeration and to grasp the quantity of sets at a glance. We know that up to a set of 5 elements, we can well learn to simultaneously capture the amount without having to count each element on its own – an important capability.

It is also important to learn that (and how) numbers can be decomposed, for instance 7 = 1+6 = 2+5 = 2+2+3 and so on. Only then are we able to calculate with numbers bigger than 10 later on.

Note: You may also let the children try to find out by themselves how to construct a house for Number 6 with the given material…

It is great if the windows are not just painted on the houses but consist of pegs. Children very much enjoy putting the pegs into the drilled window openings. If the houses consist of laminated cards, the children may place counters on the windows painted on the plates. The benefit is that the children have not only picked the correct house(s) but they also need to get the correct amount of pegs. Questions occur such as "How many pegs do I need? How many have I got? How many are missing?"

Of course, a house number must not be missed. Usually, it is a little wooden or paper flag on top of the house or a plaque, showing the numeral.

Apart from enjoying building the houses, the children develop their fine motor skills.

Number towers

The children build a tower (or chimney) in each garden with the number of floors corresponding to the number. A tower is again another way of depicting a number, this time a linear visualisation. The children learn that in each garden the same element will become exactly one brick higher, building a staircase. Hence, the 10-floor tower is much heavier than the 4-floor tower.

In the beginning and to focus on the amount, it might be easier if all 55 bricks are of the same color. You may also offer bricks of different colors. Then the children can explore the different ways of how a number can be composed. Again, the consideration of decomposition comes in, but in a different way: 5 = 2+3 = 1+4 = 2+2+1... Why should this be interesting for smaller ones? Because then they can understand the tale about Five, where her friends One, Two, Three, and Four are only allowed to join the birthday party after One has taken Four's hand and Two has taken Three's hand. Building three corresponding towers is a real eye opener.

You can also talk about addition and subtraction: "If we put two red brick onto the two yellow bricks, how many bricks do we get in total? What happens if we take away three bricks?" The children can compare the towers and contemplate how to build, say a 10 story tower out of the other towers.

Usually the younger the children are the more they are interested in just building towers of a certain height. When they are older (around 5) decomposition possibilities become more relevant.

Access to the gardens

Furnishing the gardens is very much beloved by the children because it is never the same. The range of possibilities and complexity is huge and gives you a good chance to bring in specific issues and topics.

The basic idea is that access is strictly limited to the number it belongs to! Obviously, for Four's garden, everything has to deal with 4. Sets of four items build the basis: The children may pick 4 counters such as buttons or bricks, 4 pens, or 4 beans and place them somehow in the garden. Thus, they experience the cardinal number aspect as well as invariance: All in a small place, the children see that it does not matter what a set consists of or how it is arranged! 4 is 4, no matter what. Any rectangle fished out of a feely bag works, too.

The more advanced approach includes typical things: A horse is allowed to Number 4's garden due to its 4 legs. Yet it also has 1 tail, 2 eyes, 2 nostrils – and it's only one horse... The children talk it over and negotiate, where to finally put the horse - this time. A boy in charge of Number 5's garden that day, definitely wanted to have the horse for his garden. He argued: "The horse has 4 legs and 1 tail so that makes 5!" Fortunately, the others agreed to this advanced finding...

Children are strongly motivated to find out, discuss, and decide where to put an object. This fosters cognition, language skills, social competence - and general knowledge: Why is the beetle in Six' garden? Wow, all beetles have six legs! Let's collect beetles outside or look them up in a book. We can also sing beetle songs or play activity games around beetles. And draw and design some! How about the uniqueness of our planet in the context of 1, the sun, each child? Pairs, opposites, or symmetries going with number 2? 3 basic colors? The 5 continents? Spiders with always 8 legs?

With their new view on things, toys, everyday situations and the surrounding in general become an area of exploration for the children. A 3-year-old, having lunch with her mother and brother, once completely out of context suddenly shouted out: "We are sitting in a triangle!" She was very enthusiastic having found that out, especially since 3 was her favorite number. Then she continued: "If Dad was here, it would be a square!" You should have seen that radiant smile on her face...

This is also where language and cognition come in: The children have (and *want!*) to *see* what relates to a number and to *express* it. Yet it is also a very good opportunity for children with poor language skills to participate: Although they might not know the word for something, they can still point out what they mean. Practical experience shows that exactly this happens and that this is why language skills benefit so well.

As you see, all kinds of material can be offered to decorate the gardens. Look for any kind of counters like buttons, marbles, pebbles, pencils, bricks, tokens, sticks, nuts, fallen leaves...

With open eyes, you and the children will enjoy finding items for the gardens everywhere, be it inside or outside, toys or all day material. Number Land is a true eye opener.

Hodgeypodgey, the Goblin, and Forgetmenot, the Fairy of Numbers

These two figures account for the magic thinking of children. Especially Hodgeypodgey, the goblin, is highly appreciated for his mischievous behaviour: Over and over again he messes around in Number Land, stealing windows or other furniture from the gardens, swapping house numbers or mats on Number Lane and so on. Children love to find out what is wrong and to correct the mistakes. The goblin is also a perfect scapegoat for own mistakes...
Luckily, there is also Forgetmenot, the Fairy of the Numbers, who can be called with a spell and who helps to get things right again:

Hodgeypodgey, what a dare,
tricks my numbers from here to there,
come to help, Forgetmenot,
chase that rascal from this lot!

It does not matter whether these figures appear as puppets, as disguised persons (you or a child) – or remain imaginary. This depends on what you feel most comfortable with. What works here, is the idea.

Number Tales

Children love tales and we want to give them the opportunity to enjoy the numbers also from this angle. This book provides you with specific tales about the numbers 1 to 10 that relate directly to Number Land. For example, to cure the sick Four, a tea is made out of four bits of four different herbs out of her square herb garden. After four minutes brewing time, four spoons of honey are added and stirred four times - curing immediately...
The children can listen, talk about the tales, do role-plays, develop own ideas, and so on. Interesting details for the general knowledge are embedded in the tales and, possibly, picked up later on.
You will also find many existing tales that relate to a number, such as "The three bears". Or how about making up your own number tale?

Number Songs

Music and singing are of highest value for a child's body, mind and soul. We composed little songs for each number that can go with a visit to a specific number. Their beats and number of notes correspond to the number as well as the content does to our Number Tales.
There are also two refrains that may frame a Number Land visit: One for the beginning, when the children enter Number Land, the other for the end when leaving.
As an alternative or in addition, popular nursery rhymes or songs can be included as well. Give it a thought and you will find many songs relating to a specific number! Just a few examples: "Jack and Jill" for number 2, "Five Little Ducks", "Eensty Weensty Spider" for 8, "There were ten in the bed" and so on. We list some songs in part 3.
Consider depicting a beat as a pattern on the floor, say the ¾ beat of a waltz like in Three's song. This could be 1 blue brick, 2 red bricks, 1 blue brick, 2 red bricks and so on. Invite the children to clap their hands on 1 and then slap the thighs twice and so on. Listen to Three's song again and accompany it. Try again with other songs, find other ways to accompany them also with dancing. You may know the V-step from Aerobics: It fits perfectly to Four's song...

My Book of Numbers

"Let's visit Number Land" can include many areas of creative work. A folder for collected works going along with the project is highly treasured by the children and is a good documentation of a child's development.

You may offer pictures to color going with each number. Children love pictures of their Number Land puppets and even "grumpy painters" can be motivated and will vigorously and carefully create their numbers, unknowingly training their fine motor skills... You are free to include other pictures or creative work that is part of your daily routine – but find a connection to Number Land and the children will love it! Just a few suggestions:

* Outline the number puppets on a page and encourage the children to color and to decorate it.
* Let the children glue scraps of paper, grains, or other things on a numeral on a sheet of paper.
* Numeral Rubbings: Cut numerals out of cardboard and allow the children to make crayon rubbings of them.
* The children draw pictures related to what they experienced in Number Land or what they connect to a number: A portrait of themselves (uniqueness in the context of 1), pairs of things or a picture of two friends (2), a rainbow of six colors or a beetle...
* Make shape stamps of raw potatoes and let the children stamp colorful pictures.

It is also great to have an extra sheet in each folder on which the visits to the numbers can be ticked at the end, maybe with a pretty stamp. Make sure that the folders are always available to the children so that they can work in it whenever they want in their free time.

Creativity and active games

Creative activities and active games foster a child in many important ways but often children avoid certain things because they have difficulties carrying them out. We see many such children who suddenly paint, cut, sing, run or whatever – only because the activity is related to their beloved Number Land...
Not only do the children profit in their gross and fine motor skills, social and musical skills, concentration, cognition and creativity. They also will relate these activities to their association with mathematics and thus contribute to their multifaceted picture and positive attitude.
Each visit to Number Land can be the basis for lots of activities around the number of the week: Fun games, related songs or rhymes, designing or painting, exploring specific questions, sports or outdoor activities...

Just draw on your abundant resources and relate or customize them to Number Land or a specific topic. That may be numerals made out of play dough or grainy numerals drawn with the fingers in shallow boxes or cookie sheets filled with sand. It can also be making a spider (8 legs) or a beetle (6 legs), dancing to number songs and so on.

Huge number posters for the wall document the progress of the journey: On a big sheet of paper outline the number of the week and let the children decorate the number and the space around it. Number 1 may be decorated with portraits of each (unique) child, Number 5 with prints of hands in five different colors and so on.

Part II "Let's visit Number Land"

Planning the project

Who?

Due to its flexibility, the age bracket in which children can benefit from Number Land is very wide, ranging from 3 years to 6 years – or even older children with special needs. Experience shows that already two-year-olds can playfully explore numbers 1 to 3. Children get most out of it when making their first (focused) experiences with Number Land at the age of 4 to 5 and can afterwards playfully deepen their knowledge. It shows that at that age the interest in numbers and sets starts to evolve. Children who are interested in questions such as: "How old am I?" "How many days until my birthday?" "Who is taller?" are ready for a Number Land journey.

Yet if you have children younger than 3, let them participate by just playing with the material or have older children explain things to them – provided the interest is given. If possible, you may the Number 1, see if they show interest and then perhaps move on up to number 3!

Consider how many children you have who may profit from Number Land and how you are organized staff-wise. Can you pull out to work with a separate group? This is not indispensable, you can also work with the entire group. Consider a first journey to Number Land this year and let the same children repeat it next year. The children could travel from 1 to 5 first and up to 10 the following year/semester. But many children manage 1 to 10 straight away.

How many?

The experiences are the most intense in a fairly homogeneous age group size of 8 to 10 children who meet regularly to attend to a specific number. However, it is also possible to visit Number Land with a bigger group respectively an entire class. Then tasks are being split so that some children have to look for things to be put into the gardens while others are drawing something related to the topic. It is always fascinating to see how well the children manage to arrange the things to be done with each other. There are also good experiences with Number Land during circle time with the entire group. Then the number of the week is introduced to all and specific activities are being done within other curriculum units.

How often and how long?

About 45 to 60 minutes once a week is a good duration for a trip to Number Land but experience with your specific group will show whether this is too long or too short. This also, again, depends on their age and the children's specific constitution. For 4 to 5 year olds 30 to 40 minutes may be good but we have seen many children of that age easily managing 60 minutes because they were so deeply absorbed.

Usually you can manage one number per week when taking it up during the daily routine as well. This is particularly suitable for the year before entering school. Practice shows, however, that usually there are so many things to explore and to do around a number that 2 to 3 weeks are very much enjoyed by the children. Due to that we recommend to address the 4 to 5 year olds. Do not worry but keep in mind that you can wrap so many activities around Number Land that you would do anyway to foster the children. Note that as of number 6 it may become more difficult for some children, so you may need to allow more time. Or you postpone those numbers.

Where?

It is ideal (but not compulsory) that you have a separate empty room for the Number Land visits with enough room to build Number Town, Number Lane and to play active games. Note that after each visit the material is stowed away so you do not need to block a room for the entire duration of the project. The garden of the number of the week may, however, also be placed on a table for the week with the opportunity for the children to play with it.

Each visit to Number Land can and should be the basis for lots of activities around the specific "number of the week": Fun games, related songs or rhymes, designing or painting, exploring specific questions and topics, sports or outdoor activities... These activities can take place anywhere.

This way, the children get the most out of it since they not only approach the numbers from different angles again and again, giving that multifaceted picture we talked about earlier. They also benefit in terms of creativity, musicality, motor, social and emotional skills.

A math corner

Ideally you have a math corner or a shelf with all Number Land and other material related to math that children have access to in free play during recess. This includes counters, coins, bricks, shapes, tape measures, rulers, clocks, scales, weights etc. as well as board games.

The material

One part of the material are the core elements of Number Land, the other part is all day material such as all kind of counters, washers, pennies, buttons, but also toys, pebbles, sticks, fallen leaves. Also look for specific things you want to bring in with specific numbers such as a globe to represent our earth's uniqueness, a toy horse, a spider to trigger off that topic and so on.

For Number Land itself, you need a number lane, houses, number puppets, gardens, towers, folders. It is completely up to you what material you use – as long as the relevant principles are being followed. It can be a high value of itself to construct the material together with the children. There are many ways of how to carry out the various elements and in this section we want to give you some ideas. However, for an instant start, high quality and very appealing material can also be bought (see appendix).

You may consider a big version for the floor and/or a smaller version for open play/ a math corner.

Number Lane 0 to 20

Slip resistant material such as Linoleum, carpet pieces, wooden plates of size 12 in. by 17 in. approximately. Make sure that material and colors are uniform and calm but let 0, 5, 10, 15, and 20 differ from the other numbers to make them stand out. You may additionally paint two feet or "start" on 0.

If you have stairs, decorate and use them as a number line as well.

Houses

You need houses with 1, 2, 3, 4, 5, 5+1, 5+2, 5+3, 5+4, 5+5 windows. That makes houses with 1 to 4 windows twice and seven houses with 5 windows.

You may use wooden bricks, cut a wooden beam, cardboard boxes, or plastic boxes. A good size for the floor version is approx. 6x6 in., for a table version about 2,5x2,5 in.

Ideally holes for the windows are drilled so that the children can plug in windows such as cork, sawed wooden rods, pegs or wooden plugs. You may also find caps that you can use. Or you nail the windows and have magnets ready to be put on.

For the small version pegs from a game will do or dowels.

For the house numbers you can paint a little card. To make flags, paper tags on twigs, a straw, or a wooden stick are nice.

Number puppets

You may choose the material according to your preferences and those of the children. Hence a proper size for small hands is important, room to store them and perhaps wash ability. Good sizes vary from about 8 to 18 inches. Most importantly, the puppets should have a face (to look into the correct direction) and details on them that are typical for the number, such as an amount of buttons, stripes and so on.

Small children often love big cuddly numbers made of soft fabric or terry cloth with a soft filling.

Highly appreciated are also robust wooden numbers that can be painted. Little arms and legs of thread and big pearls look vivid and make funny noises.

Alternatives are paper -mâché or (not so durable) cardboard paper. Or you paint the number puppets on an ordinary sheet of paper, laminate them and cut them out. Perhaps you find ready-made numbers to decorate.

Number gardens

Geometrically shaped gardens are the most impressive but for the beginning you may also use hoops. With a little effort, you can construct great gardens out of green outdoor rugs from the local home improvement store. This looks very natural, you can cut it easily and it is very durable. Canvas will do as well as linoleum.

Basis is the circle out of which you can construct each geometric shape (except the ellipse for Number Two). Evenly increase the gardens to visualize again that 1 is a much smaller number than 10. For the floor version we suggest 25 inches diameter for One's garden up to about 28 inches diameter for Ten. For a small table version, consider 4 to 8 inches. You may use a regular compass for the table versions. For the floor version, tie a nail onto the one end of a thread and a pencil onto the other end. That will serve as a big compass. Place the nail into the center of the circle and draw the circle (The thread must be 12,5 inches long if you want 25 inches diameter).

We suggest that for the polygons 3 to 10 you make patterns for each garden on pieces of paper by using a regular compass and a protractor. These patterns can be reused for any big and small gardens to come! Now we do a little math:
Each circle has 360Õ. Divide these 360Õ by the number of corners (vertices) you need and you get the specific angle to construct the shape. For the triangle we calculate 360Õ divided by 3 = 120Õ. Draw a first line, say 3 inches long, from the center of the circle. Place the middle of the protractor (where it shows 0Õ) on the circle center. Now measure 120Õaround the circle center from the first line and draw a second line. From this line, measure 120 Õagain and this will give you a third line.
This pattern is the basis for any triangle you may want to construct: Place the pattern onto the outdoor rug. Draw a circle with the desired radius around the center of the pattern. Extend the lines from the pattern and mark where they cross the circle to define the corners of the triangle. Draw lines between the corners, cut along them and you get a triangle of the desired size.
Do the same for the other polygons. The other angles are: square 90Õ, pentagon 62Õ, hexagon 60Õ, septagon 51,4Õ, octagon 45Õ, nonagon 40Õ, decagon 36Õ).
An ellipse is constructed around two points – giving an impressive explanation how Number Two's garden relates to that geometric shape! You need a string of about 22 inches, 2 nails tied to each end, a pen – and a helping hand. Let one of you hold the two nails onto the outdoor rug, about 12 inches apart so that the string lies loosely between the nails. Your partner moves the pencil along the thread and draws the upper bow. Then she does the same to the other side and draws the lower bow – and out comes a beautiful ellipse!

Number towers

You need 55 bricks of the same shape to build number towers 1 to 10. Suitable are all kind of wooden or plastic bricks usually at hand in great quantities anyway. It helps if you have at least two basic colors so that the children can explore the decomposability of numbers: Five's tower can be built with 5 red bricks, but 2 red + 3 green are also 5 and so are 1 red + 1 yellow + 2 blue + 1 green!

You may also use yellow, blue, red and green stickers on existing bricks, or paint a dot onto each side so that the children can explore the bricks by rotating them.

Number caterpillars

These little fellows are a nice, beloved complement for the gardens: Make beads for the number 1 to 10 looking like little caterpillars. Each caterpillar consists of as many pearls as suitable for the number plus an extra, bigger pearl for the head where you might paint a face. Let the first five beads have pearls all of the same color, for example red. As of caterpillar 6, add one different pearl so that you get 5 red plus 1 blue pearl. Go on until you finally have the tenth caterpillar with 5 red pearls followed by 5 blue pearls.

Counting board

Another helpful item for the gardens are 10 little cards, all with two lines of 5 dots next to each other. The children place them into each garden and count the respective amount of counters on the dots. Offer two versions of counters and let the children experiment how they can arrange the counters on the card.

Prelude and the very first visit

Getting started the very first time is the hardest part of a Number Land project. As soon as your and the children have made first experiences and are familiar with the material and the elements you will know how to handle it. Encourage the children to bring in own ideas and let them participate in the way the project runs.
It is a good idea to introduce the children to their upcoming trips to Number Land to make them curious and focused. Ideas are a dime a dozen! One idea is that the children „find" One who starts a conversation about numbers:
What numbers do we already know, where can we find them, what can we count. Finally, it invites the children to visit it in Number Land and tells about Number Town, her neighbours, Hodgeypodgey, Forgetmenot. You may also read out the introducing tale in this book.

On the first visit to Number Land, getting to know the elements and the things to do takes a big part. Organize the houses, windows, house numbers, and towers on a table. On another table put counters and typical things for number 1 such as: a globe, a unicorn, a picture of the sun transporting the idea of uniqueness. Number Lane is already laid out completely up to the number you want to reach in the end.

Consider whether you only lay out One's garden and have the other gardens added one by one when visiting the other numbers. This is logical but the room might look a little bit empty. Alternatively, you may as well lay out the garden 1 to 5 at once and let the children find out which one might be One's garden and then focus on garden 1. The advantage is that the gardens 1 to 5 differ most in their shape and the empty gardens give a good picture of the idea. Whenever you are done with number 1 you move on to number 2, focus on that garden and activities and so on. When it comes to number 6, add garden 6, then 7 and so on up to 10. A third option, especially with older children, is to lay out all 10 gardens at once. This gives an idea of entire Number Town and piques curiosity for things to come.

Accompanied by the first song, the children enter the Number Land room and have a first look around. Tell them that Number Lane leads the way to Number Town and it is obviously important to learn how to manage Number Lane in order to find their way! Invite the children to have a first try and to walk over Number Lane one after another, if possible counting out loud at the same time. Watch the children carefully to see how they manage. If someone is too shy, take his hand and walk together but do not rush a child. Some children prefer to stand watch from the distance first. The children place one counter on 1 on Number Lane.

One may welcome the children and explain what to do. It can ask questions such as: "Do you find my garden? My house?" Gather around its garden, talk about number 1 and have one or two children look for the correct the house, then someone else get the window, someone the house number and so on. Decorate the garden together then talk about what you see. Make the children detect what they have only one time (nose, head, mouth) and that each of them is unique, too! One may thank the children for the beautiful garden by singing out its special song to them. Sitting in a circle, you may then read out One's tale. It is nice and impressive if you accompany that with a little role-play with the number puppet and a few props.

That will probably more than do for the first visit. Do not overload but let the children determine the pace and how much is done. They have so much new to take in that it might even be advisable, for instance, to read out the tale another time.

Hand out the Books of Numbers and tell the children how to use them, then stamp it. To the farewell chorus the children say goodbye to One and leave Number Land. Maybe they already manage to walk backwards on Number Lane?

Suggested schedule for further visits

Always remember that you are free in your organisation and that the children determine a lot themselves. Keep in mind that rituals and routines are important, too, to give orientation and help to concentrate. Prepare a schedule for each visit but allow changes according to the reactions of the children or what they bring up.
It is of utmost value for the children if they can explore and discuss on their own, especially when it comes to furnishing the gardens. Children usually have a "protective shield" against things that are yet too complex for them. Within the same kindergarten we one year experienced a group that was very advanced and enjoyed all elements a lot. The group of the following year was much weaker. Those children by themselves focused on furnishing the gardens and playing on Number Lane but were not interested in the tales.
Prepare a folder in which all teachers involved collect good ideas and activities over the time. You can also document experiences there.

Preparation

Lay out Number Lane as the path to Number Land. When the children are quite familiar with Number Lane, dislocate the cards more or less.
Lay out the mats for the gardens. Place them in the correct order or on a pile if you want the children to do the job. Remember that complete Number Town builds a full circle. So if you lay out only 5 gardens they build only the first half of that circle.
Place the houses, window, numbers, puppets and the bricks for the towers ready.
Arrange counters, toys and other material for the gardens on a table. Material that you specifically want to use for today's number, keep on a separate table and cover it. If you want to keep it simple, offer only counters. For a more advanced approach or to tick off certain topics include typical or ambiguous things.
Other material: Props for the tales, activity games, songs, coloring pictures and stamp for the Book of Numbers…

Main part

The children enter the room to the chorus of the respective number.
The children count tokens on Number Lane up to today's number. Perhaps Hodgeypodgey has messed up Number Lane! Then the children call the fairy with the spell, rearrange Number Lane and walk it for control.
The children organize all gardens they already have visited. Lining up on Number Lane helps to determine who wants to do which garden. Each team coordinates who does what by itself. The children choose a) similar things according to the number or b) something typical for that number. The children may also bring things themselves and argue why they fit a specific number!
A nice idea is to have a "guard" for each garden: The other children bring things and explain why they offer them but the guard may decide whether or not he agrees.
When all gardens are done, check each garden together, express compliments and clarify questions.
Do the new garden of the number of the week together, maybe in a dialogue with the number of the week. Here, you can introduce things relating to the number. Allow enough time for this core activity and let the children contemplate and debate.
The number puppet brings in her song and/or her tale. Listen, move, make music to the song or another related song. Read out the tale, possibly with the help of props and with involvement of the children.
Pick up topics of the tale, build activity games around the number of the week, for example a seasons game going with number 4 or one with pairs in the context of number 2.

End

Hand out coloring pictures and initiate ideas for Book of Numbers activities. Stamp the table with the visits to Number Land. Play the farewell - chorus and let the children move out of Number Land, possibly backwards over Number Lane.

Throughout the week

If some things had to be left out during the visit you can pick it up during the week. Building Number Town and activities on Number Lane always take the biggest part so maybe time was too short for the tale, the number song, a game or a topic related to a number. That is no problem because you may elaborate on these topics later in the week and/or in circle time.
Some ideas: In circle time, sing related songs and nursery rhymes, pick up and deepen topics. In sports, do a relay with x elements. Take a walk and find and collect material for Number Land. At lunch offer x pieces of fruit. Relate creative work to the project: Paint pictures with x colors or things, make a collage with x different material or colors, design a big poster for the number.

Ideas for the beginning of a Number Land visit

Presents: We clarify what number we want to visit today. Maybe the children would like to bring a present from the room, from outside, from home. What would the number like? The children collect / create presents according to their ideas and bring them along.
The key: We knock on the door according to the number of the week before we can open it.
Where is our host? Is it hiding? Let's call and look for it!
Expert talk: The new number discusses with the children the previous numbers before building the new garden together.

Activities and games on Number Lane

This is a collection of proven activities on Number Lane but feel free to vary and to add own ideas. They are suitable for Number Lanes of any length. As an alternative to the cards and for a change, the children may want to paint a Number Lane outside with chalk. A staircase is also a great Number Lane!

The children walk forward and count out loud with each step. Later this can also be done backwards. You may show the way or accompany unsure or shy children. Another child could also act as a guide. Keep in mind that some children may want to watch first, so do not rush them.

Do they manage Number Lane with closed eyes? This is a first concentration task.

For a change, the children go/hop Number Lane in two or three steps at a time. Where do they get? The children interrupt their walk over Number Lane: Which number is next? Which one came before? Can they move on and count forward once they have stopped their counting routine? This requires that the child did memorize the previous number and knows the order.

Number Lane is in correct order but one number is turned upside down (later on several or even all). Can the children walk and count properly although they do not get all the numerals?

A child rolls a dice (or draws a card) and moves forward (or backward) accordingly. Where does he get? The other children guess, too, and may hold a card with the correct number up or show it with their fingers.

All cards except 0 are turned upside down so the numerals are hidden. A child rolls a dice (or draws a card) and moves forward/backward accordingly. The child stops and tells the number he reached. Then he turns the number mat. This goes on until all numbers have been revealed again.

The children decorate Number Lane up to the number of the week: one counter on 1, two on 2 and so on. They may the line up the counters next to the numbers to get the picture of a staircase. It is fun when a toy lorry unloads pebbles at each number or little flowers that can later be "planted" into the gardens.

Hodgeypodgey has messed up Number Lane! Will the children manage all the same? Here are some of his tricks: Incorrect order of the numbers; some cards are turned upside down; some are missing / hidden; the cards are spread out over the entire room...

The cards are spread out over the entire room. Do the children find their way in the correct order? Can they hop from number to number if they are not too far away? Also if they always count two more, three more? Variation: The fairy (or someone else) calls out a number and the children run to the correct card.

"Toy Race": A child rolls a dice, moves the respective number of steps forward and lays down a toy on the card. Then it is up to the next child to start from there. How long will it take to reach a specific number? Note: A two-sided counter may be brought in to decide whether to move forward or backward! Depending on the length of Number Lane it may be a good idea to prepare the dice so that you can only get 1 to 3. As an alternative, let the children draw cards from a little bag.
Alternative: This can also be played with two or more toys, making a little competition. Who will reach number 10 (or 20) first? Or: Each child has a toy and as soon as they have reached the goal number they receive a stamp or their hand or something else little.

"Treasure of Gold": This very popular game is most fun on Number Lane 0 to 20 and a good way to train the number symbols beyond 10. However, 0 to 10 is also possible, but then prepare a dice with only 1 to 3 dots on the sides.
Two teams of robbers fight over a treasure of gold (a box or a bag with some little gifts in it) that is placed in the middle of Number Lane. Number 0 and number 20 (10) are the robbers' dens into which each team tries to drag the treasure. Allot who starts, for example Team 0. They throw the dice and move the treasure towards their den. Example: They throw 3 so the treasure moves 10-3=7. They place the treasure here and Team 20 is next. They throw 5 so the treasure moves 7+5=12 and is placed on that card. Then it is Team 0's turn again and so on – until one of the team has the treasure in its den. A good alternative (to shorten the game) is to have each team member throw the dice one time. Will the treasure end up closer to Team 0's den or to Team 20's den? That team wins and is allowed to raid the treasure! But it is always nice if the other team may pitch into the leftovers...

Ideas for the Number Towers

- Build a tower in each garden, then put them next to each other and see the stairs.
- Use expressions such as higher / lower, bigger / smaller, double/half, heavier / lighter, next to…
- The children build towers next to each other that are of the same height but differ in their color configuration: 5=1+1+1+1+1=2+3=2+2+1…
- Build a tower, count the bricks then throw it over. How many bricks do we have now? Still the same, although the bricks are in a different order!
- How can I turn a tower of 5 into a tower of 3? What do I do?
- Double/halve towers.

Some ideas applicable for all numbers

You can bring in all proven activities you have – just relate them to the number of the week and the children will get the most out of it! Here are just a few ideas to start with. Many more are suggested in Part III.

Listen: How often do you (or the Fairy of the Numbers) gently hit a drum or a triangle? The children quietly go to the respective garden or hold up a card or something else.

Town construction: Build a town with x towers – of bricks, cubes, paper mâché...

Where is the number? Draw all kinds of numbers on a piece of paper. Let the children find the number of the week. Shapes are also great.

Into the middle! On call the children reach out as many arms as needed to build the number. You may also use legs, heads, nose…

Laying geometric shapes: Rebuild the gardens with different material such as paper stripes, pens, straws, the children themselves. Long ropes are fun to walk on. Later on, the children may sit inside and listen to the tale. How about being the corner of a garden?

Construct really huge geometric shapes on the floor together with the children. A long rope with some chalk is great for a circle or an ellipse. For the other shapes, the children are the corners and arrange ropes, sticks, paper stripes or their arms until they get the desired shape. Let the children walk on the shapes, lie next to them, form shapes with their bodies on the floor. In little, offer pens, straws or tooth picks and play dough so the children can construct little shapes themselves.

Feely bags with counters, pebbles, marbles are great: The children reach into the bag to collect a specific amount. Or fill feely bags with different quantities and let the children feel how many are inside, compare. Shapes are also great for a feely bag, or consider for example different quadragons in the context of 4!

The grand finale: a Number Land Party

For the ending of a Number Land project a Number Land Party is a great finale. It can take place after having finished Number 5 (who in the tale celebrates her birthday anyway) or after having finished with all Numbers 1 to 10. There are many ways of organizing it – from very little and intimate to really big.

A nice little version is to gather for a last time, play a few games (Treasure hunt on Number Lane 0 to 20 is a good one) and to offer a “number buffet”. That may be offering some cheese cubes and grapes the children may put on small skewers, a salad out of x different fruits, a cake shaped as a numeral, shaped cookies and so on.

A huge party can include the entire institution and the parents. The children can present some of their work, give a role-play, dance and sing… Offer different tasks at stations and stamp tickets for each station.

A Number Land certificate and/or a little gift for each child can be handed over on that occasion.

Part III

Tales, songs and activities about the numbers

Welcome to Number Land

Lovely Number Land reaches from the Sea of Infinity to the Mountains of Multiplication. In between are so many different countries that I am not able to even name them all. Some of them though I would like to mention, in case you would ever like to visit Number Land.

The Sea of Infinity has swept endless grains of sand to shore which, in the course of billions and billions of years, has turned into a wonderful place where there is much to be seen and to be discovered. There you may erect sand castles and dig deep holes. Collectors among you will find, hidden in the sand, colorful clamshells and every once in a while even a stranded sea star.

But be careful! For behind the dunes stretches the Desert of Forgetfulness! Visitors who venture too far into this desert will be forgetting all numbers they have learned previously in Number Land, and shall never find their way back! Except when they meet the good fairy Forgetmenot. She would then take the lost wanderers by the hand and show them the road back to safety.

There is a wonderful avenue leading from the golden shore to a pretty little town. Its name is Number Town, because all our numbers are living there. Each number possesses a nice house along the Number Lane, that is the Promenade of Numbers, with a garden around it and a chimney at its top. You will, in due time, learn much about these numbers, about their lives, and the exciting adventures.

In the West, behind the walls of the town, spreads the Forest of Mistakes with its ancient trees. Nobody dares to venture into the Forest of Mistakes without having befriended one or more numbers in advance, because in the center of this forest cheeky Hodgeypodgey, the goblin of Numbers, has his domicile!

Hodgeypodgey sports a long pointed nose, which he likes to stick into other people´s business, a big magic hat on his head atop of tousled hair and his clothes are all torn. But he does not care because he loves disorder. If he is not somewhere else, teasing poor numbers, he is sitting in his broken-down, dilapidated hut slurping cold coffee out of a cup with a broken handle and is on the look-out for wanderers going by to mess them up.

Luckily I do happen to know a spell, which, in such a dangerous situation, will call upon the good fairy Forgetmenot. Do you want to hear it? You just may need it sometime!
It goes like this:

Hodgeypodgey, what a dare,
tricks my numbers from here to there,
come to help, Forgetmenot,
chase that rascal from this lot!

Have you by now memorized the spell?
East of Number Land are wide green meadows and numberfields. There the numbers grow their vegetables and salads. On a bright and colorful meadow, surrounded by fruit trees, lives the good Fairy of Numbers, Forgetmenot. On nice warm days in spring and summer she puts up a tent of fine and colorful fabric as her living quarters. When, in fall and winter, the weather turns cold, she uses a spell and – hey-presto! – she becomes teeny-weeny and moves into a snugly little dwelling in a cherry tree, created long time ago by a stroke of lightning. From there she keeps watch over Numberland and looks out for naughty Hodgeypodgey to prevent him from doing too much mischief.
The Fairy of Numbers is a real beauty, with eyes that are the colour of forgetmenots and a flowing cape over her shoulders, embroidered with all the numbers of this world. With her glittering magic wand she helps the numbers in Numbertown and brings Hodgeypodgey´s mischief to a good end. And if you will ask her she will gladly help you.
You do remember the wording of the spell? Or have you forgotten it? Has, by any chance, Hodgeypodgey smuggled himself in among us and created a hodgepodge? Hopefully not!
Looming up behind the Forest of Mistakes and the Fields of Numbers are the Mountains of Multiplication Tables. They are higher than the highest sky-scrapers of the world. It takes plenty of effort to climb them but if you are successful and reach the top you will be most richly rewarded: You will see beautiful Number Land with its golden shore, the Desert, the Forest, the Meadows and Number Town spreading at your feet. And everything is looking so magnificent.
But for now I will quit telling you more. I am pretty sure you are all very clever persons and intelligent people and have well memorized the spell I have taught you. You will surely need it now when entering beautiful Number Land.

One - The Uni-No-Horn

There is a nice little town in Number Land. The numbers are living there in their small houses. Recently I was taking my first walk along the Promenade of Numbers and had, all of a sudden, barely time and the chance to jump to the side. I was almost hit by a biker. "Ding-a-ling!" rang the bell of a bike, with a short, little man riding it. He waved a shimmering white object towards me and sped along. "Such a nerve!" was my reaction. "To scare innocent people on the sidewalk!" Then I passed the first house. Listen! Wasn´t there somebody crying? Yes, indeed! A little number was sitting in the garden and crying. "Whaaah – Whaaah!"

"Can I help you?" I asked.

The little number looked at me with tearful eyes and said sobbing: "That goblin, that rascally Hodgeypodgey!"

"What has he done to you?" I asked.

"My poor little creature", sobbed the little number, "can´t you see what this monster has done to it?" and pointed to the other side of the garden.

And what do you think I saw there? A little white horse was standing there in the grass - nice to look at with its white tail and glittering, shimmering mane.

"A nice little horse", said I, "but shouldn´t we introduce one another? My name is....... And what is yours?"

"I am One", sobbed One, "and can´t you understand?"

"But your little horse looks perfect!"

"That´s exactly the trouble!" said One. "Don´t you think a Unicorn should be having a horn?"

I took a better look at the little horse. On its forehead a spot marked the place where formerly its horn had been. Poor little creature! Considering this, one must come to the conclusion that a Unicorn with missing horn becomes a No-Horn!

"Now you´ve got it", lamented One, "that miserable Hodgeypodgey!"

Hearing this I suddenly remembered that the near midget, who almost had run me over with his bike, had been waving a white object. The pixy had stolen Unicorn´s corn and was now over the hills. But how could I now be of help to both One and Unicorn? A wheelbarrow was leaning against the wall of the house.
"Take a seat in this wheelbarrow", said I to One, "and we go after the pixy."
"I don´t dare", said One anxiously, "I´m so small."
"Don´t you worry", said I, "I am with you!" And that was it. One jumped into the wheelbarrow and off we went. Through the town and its gate to the Forest of Mistakes.
"I´m not going into there", wailed One.
"Again, don´t you worry", said I. "Just close your eyes and I bring you to the pixie´s hut."
One nodded bravely. We hurried through the dark forest as fast as we could and arrived within one hour´s time at Hodgeypodgey´s hut.
"You may now open your eyes", said I in a whisper to One, "we have arrived."
"Oh my goodness, what a mess", whispered One in turn, after having taken a look around. "Everything is full of rubbish and the hut is so dipilated it´s a miracle that it´s still standing. What do we do now?"
"I´ll get some reinforcement", whispered I in turn and then cited the spell:

Hodgeypodgey what a dare,
tricks my numbers from here to there,
come to help, Forgetmenot,
chase that rascal from this lot!

Just as I had finished, a light appeared before my eyes and I had to close them. After a while I could open them again. I saw the good Fairy of Numbers standing in front of me.
"You have called me?" she asked. And then One told her the whole story.
"This is too much!" exclaimed Forgetmenot indignantly. "Hodgeypodgey you come out at once!"
The door of the hut creaked open and Hudgeypodgey looked at us distrustful. "Just what is it that you want from me?" he played the innocent.
"You know exactly what we want!" snapped the fairy. "Return that corn immediately to its rightful owner, you thief!"
"And what do I get in turn?" asked Hodgeypodgey insolently.
Forgetmenot just looked at him. Finally she said: "Listen, both of you! I will ask you one question. And one question only. Him who gives the right answer gets the corn. Understand? All right: What is more than none?"
The pixy did some thinking while scratching his chin. Finally he said: "I´ve no idea. Who should know such a thing?"
"I do", exulted One joyfully, "that is me! I am One, more than none! Yippee!"
"The corn is yours", said Forgetmenot smiling and disappeared in a flash.
Hodgeypodgey returned the corn to One, poor No-Corn became again Unicorn and I just came hurrying to you to tell you the happy end of the story.

In Land One

One is my name, you´ll know me soon
my little song has just one tone
live in the first house on the shore
with number One on my door
carry in my face a quite big nose
who ever met me will keep me quite close

Setting up of Gardens of Numbers.

At their first visit children set up gardens up to the number five only.
Who or what may come into One´s garden?
Which objects are characteristical for One-Land?
One hammer, one magic wand, one picture of a unicorn, one monocle...

Introduction of Number Lane

Number Lane is an especially fine way of introducing to children the world of numbers with all its kinaesthetic experiences.The only way into Number Land is Number Lane. It is appropriate to start the discoveries in Number Land with practical works on Number Lane. The most important activity on Number Lane is walking it and counting, eather aloud or silently. By walking forward, the numbers become larger. Each step forward means an addition of the number one. The same applies to 2, 3 or more steps. Thus cognitive results are being supported by motion. In the beginning children walk Number Lane forward and count aloud in the rhythm of each taken step. This is for many a child quite complicated. Is the walking forward and counting aloud successful, it is advisable to do the same backwards. It is a simple but pregnant method, and children should actually walk backwards. One may depart from Number Land, at the end of the project, thus.

Ideas to construct a Number Lane

A Number Lane can be constructed best together with the children´s assistance. Numbers 1 to 10 will be painted with water colors on conveniently sized sheets of paper and placed and fastened in the right sequence on the floor. It is most important to make sure that children won´t slip and fall on the arrangement! The constructions can differ as to the ideas of the constructor. In the open, numbers can be simply drawn, or painted, or chalked on the asphalt. Stariways can also be used and are most convenient for counting.

Make a score in the bucket

The children form two teams and throw alternately a ball into a bucket, placed conveniently on the floor. Each child throws the ball only once and each score merits exactly one point, resp. one object (one marble or such). The number of throws for each team must be fixed in advance. The team with the highest score is winner.
To check the score, it is not necessarily important to be able to count: It´ll suffice to compare the scores one by one (i.e. place the objects next to one another).

Kerchoo and gone

Number One has a terrible cold and is constantly sneezing, which is a very dangerous thing to be done in Number Land. Because each time number One is sneezing, one item is disappearing in Number Land! Number One is sneezing and an adult takes one item from Number Town and puts it away. This is repeated so often, as the children catch the idea and tidy things up by themselves. At the end One sneezes for the last time and disappears itself. The number is now at home again in Number Land.

Each one of us is unique.

What would the world look like if some things were to be found existing in strictly one form only? Children find it more than funny to create pictures in that way or to look at such prepared pictures. A picture of a face, having just one nose and one mouth is nothing special. But what about a face with just one eye, one ear, one tooth or one hair? Or a bicycle with just one wheel? Well, that would be a one - wheel-bicycle. With such settings of a task (for example to hop on one leg), children can be sensitized for the uniqueness of some forms and things.
Which forms are actually to be found only once? The sun, the moon, my mother, my father and finally myself.
I am unique.

Activities during the week

+Number Breakfast : Is number One the number of the week, each child receives, (for example) one apple or one orange.
+Start a Book of Numbers into which the children can draw and paint their respective pictures: singular pictures like the moon, face with one nose or one eye, unicorn etc.
+Exercise on Number Lane: A pebble or a marble or some such convenient thing is placed on the picture of number one to symbolize the connection between angle of quantity and angle of disposition.
+A song matching singleness.
+Counting some things, like stairs etc.
+Inventing and telling stories: Imagine, last night I dreamt I was visiting One at its home. Each thing in its house exists only once. One offered me a drink, I took one mouthful and the glass was empty...

Two - Rescue-rescue from-from tree-tree

Yesterday I took another a walk along the Promenade of Numbers. The sun was shining, birds were singing in the trees and a cat darted across the road. Then, all of a sudden, somebody called: "Hello, hello!" I turned around but nobody was there. Just as I wanted to go I heard again the voice: "I I am am up up here here!" I looked up but there was only the sun dazzling my eyes and two high trees.
"Anybody around?" I asked.
"Yes yes I I am am up up here here."
"The poor chap is stuttering like an old engine", said I quietly to myself.
"I I am am not not stuttering stuttering at at all all!" called the voice angrily. "I I am am Two Two and and speak speak always always like like this this."
"Forgive me please", I replied, "but I was not aware of this fact."
As I looked up again searchingly, I discovered Two: the little number was sitting high up on a thin branch and was looking unhappily down at me. "Watch out and don´t fall down!" I called.
"I I just just wanted wanted to to rescue rescue my my cat cat", said Two. "My my dog dog has has chased chased her her up up the the tree tree and and she she didn´t didn´t dare dare to to come come down down alone alone."
As you may see, it wasn´t easy to understand Two, because Two said everything twice! For simplicity´s sake I will from now on leave out her every second word, that is half of what Two was saying, or else we´ll be sitting here for the next two hours.
What Two was actually saying, was: "I just wanted to rescue my cat. My dog chased her up the tree and she didn´t dare to come down alone."
"Where is your cat?" I asked. "I can´t see her."
"She just jumped down", answered Two.
"Is that so?" I said. "I think I saw her. She has crossed my way a short time ago. Come on, climb down."

"I can´t", said Two pitifully. "I´m sitting tight and I don´t dare to move. Could you please help me?"
Two was really in a fix. "Is there somewhere a ladder to be found?" I asked.
"I don´t think so", answered Two. "Oh my, what shall I do?" said the poor number plaintively.
"Just stay put", I said somewhat worriedly. "I´ll go and get some help."
"Thank thank you you", said Two bravely.
I was thinking, what could I really do? Say, children, what would you have done in my place? How would you have tried to help Two?
I meant to return to One to ask for a ladder, when something pulled at my pants leg. I looked down and what do you think I saw? A little doggie! He placed a long stick at my foot and begged me to play with him. I threw the stick as far as I could. The doggie went after it just like lightning. I ran the other direction and arrived at number One breathlessly.
"Dear One", I said breathing heavily, "excuse my intruding, but could I borrow a ladder from you?"
One shook its head regretfully. I told the little number One about neighbour Two´s unfortunate situation.
"I have an idea!" exclaimed One. "You stand under the tree and I climb on your shoulders. This way we should be able to rescue Two."
No sooner said than done. "I am so glad", said Two standing again on firm earth. "Please would you bring me home?"
We escorted number Two to its house. Upon our arrival it started getting dark and the moon was rising.
"I am very grateful to you, both of you", said Two. "May I invite you to a double bowl of punch?"
One shook its head. "I must run, my Unicorn needs to be groomed", it said and hurried off.
I also declined the invitation. "It´s getting late and I must go home."
"Why don´t you come and visit me in two days? We shall have double-stew and double-ice! Till till then then good good bye bye", said Two and waved both hands.
After I´ve made a few steps I turned around twice.

The first time I discovered the cat, she looked out of the left window.

The second time I saw the doggie. it was barking at me from the right window.

"You rascals", said I and loughed. "You never give peace", I said laughingly.
And this was my second adventure in beautiful Number Land.

In Land Two

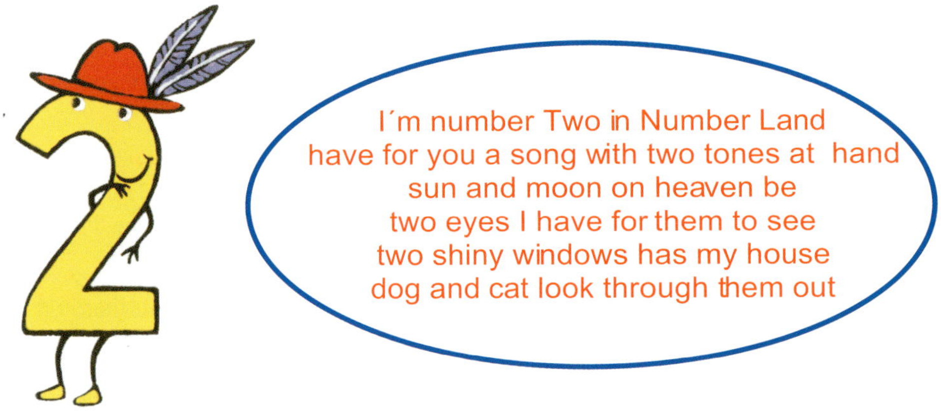

Setting up of Gardens of Numbers.

Who or what may enter Two´s garden? Children give reason why some objects or creatures may come into garden two or why not. A hen or a kangaroo? Yes: Both have two legs. A dog or a cat? No: Both have four legs. A bird? Yes: It has two wings. A bicycle? Yes: It has two wheels. Also a pair of shoes, socks or mittens are allowed. And of course twins!

Exercise on Number Lane

Number Lane should be an essential part of each discovery in Number Land. The degrees of difficulties may be increased step by step. Important! Children must be sure in walking even and regular. If they are not, this should be exercized till satisfying results show. Are the children´s steps even and regular, or in good harmony together, a rhythm of two or three or even four can be practized on Number Lane. Thus even or uneven numbers can be emphasized. If these rhythms are mastered by the children, numbers my be covered and thus not seen by the children.

Pairings

Let the children find the respective pairings: You and I, wife and husband, fork and knife etc.
Opposite: Yes and No, young and old, hot and cold, rich and poor, left and right etc.

Blot pictures

Children are folding a sheet of paper in the center and paint something with ink or water colors on one side of the folded paper. Then they fold it. The result is an axially symmetric blot picture which shows everything that has been painted doubly. Here: several blot pictures have been arranged to a blossom.

Mirror games

A small, rectangular handmirror, placed at a right angle on a picture. The picture and mirror show axially symmetric images on which everything can be seen twofold. A mirror placed at the right angle to a 3 enables the children to "conjure" an 8. This method enables us alos to search for symmetrical axis on pictures of faces or, for example, the 8.

Clap twice

All children form a circle around another child. The child in the center throws a ball to one of the children in the circle. This child must catch the ball, but only after having clapped its hands twice. This play needs plenty of exercise.

Cat and mouse

Two children are cat and mother of mice. The rest of the children are little mice who have to protect themselves from the cat, who, naturally, tries to catch them. They are doing this by holding fast to mother´s hip or shoulders or to the last little mouse in the row of mousechildren clinging to mother. The cat must try to catch the last little mouse by touching it, while the mother of mice is trying to hinder this by running away or protecting her children by swinging her arms. If successful, the child who has played the cat becomes the new mother of mice and the little mouse who has been caught becomes the new cat. The previous mother of mice is out.

Activities during the week

+Number Breakfast: Is number Two the number of the week, each child receives, for example, two tangerines.
+Book of Numbers: Draw or paint a picture of, for example, an animal with two legs or of other pairings.
+Number Lane: A pebble or a marble comes to be placed on number 1, and two of them on number 2.
+Jump from the second stair of the stairwell.
+Play pitch-penny.
+Speaking speaking as as the the number number two two.
+Invent and tell new stories.

Three - Visiting Neptune´s Realm

As you may remember, every number has a small house in Number Town. One has a Unicorn grazing in the garden. Two has two high trees in front of the door and many-many birds in them. Every once in a while the doggie chases the cat up the tree, where she tries to catch a birdie. But the birds fly away as soon as the cat climbs up the tree. Fortunately there is a garden near, where they can land safely. That garden has a house with only three corners. And, who do you think is living in that house? It´s Three!
In Three´s realm everything is triangular: the door of the garden, the three legged table and even the bathroom.
Would you like to know why the birds can land so safely in Three´s garden? Because the garden can be entered only by threes. The dog and the cat are only two together and as such they can not enter. Would they be clever enough to unite with Unicorn they could enter. Because then they would be three together: A dog, a cat and a Unicorn. But such a tough mathematical problem can be solved neither by dog nor by cat. And that is the reason, why the birds are safe in the tripplegarden.
Recently I was visiting Three. As a token I brought it three feathers, a red, a yellow and a blue one. I knocked three times at the triangular door. I sat in a three-legged chair. Three politely offered me a drink. But after having taken three sips of the terribly sweet triple-brew I had to hurry to the bathroom. There I looked into a triangular mirror. My face changed all of a sudden! A mustache was growing under my nose and I had a hat on my head with three corners with each a red, a yellow and a blue feather on it.

The ground under my feet began to shake and I looked around bewildered. I was suddenly smacked in the middle of the ocean on board of a ship with three large sails. On top of the highest mast a flag was flapping in the wind. It sported three crossed swords.
"Oh my goodness", I sighed, "I am ona corsair!"
A ragged character was approaching me. He said in a rough voice: "Hy-ho, Cap´n, welcome on board."
I simply nodded.
The mate didn´t let loose. "To sail on the Sea of Infinity, looking for treasures, is the most exciting thing in the world!"
I just nodded again.
"Your diving bell is just over here, Capt´n", said the chap.
"Oh no!" I thought. "Now I even have to dive!" But I couldn´t and wouldn´t simply admit that I wasn´t a buccaneer. The fellow would have thrown me overboard. So I donned the diving bell and jumped into the water.
Splash! – made the waves as they closed over my head. I sank deep and deeper. It took me thirty minutes to hit sea bottom. Colorful fish swam around me and a nosy little seahorse looked me over. Then an odd something approached me. Half man, half fish with a trident in his hands. He beckoned me to follow him. We stopped in front of a cave.
"Mortal", he boomed, "you have successfully reached Neptune´s Realm. The fortune you are looking for is hidden in this cave. If you are taking it, you must forever remain on the bottom of the endless sea."
I was shocked. I never and never wanted to remain in Neptune´s Realm.
"It´s your choice", boomed Neptune, for it was him personally. "But I will grant you three wishes, if you´ll leave the treasure untouched!"
"I – I´ll just have – have the three wishes", I stammered.
"Then wish, but wish well!"
"I wish", I began, "you would be my friend."
Neptune smiled at me. "Your first wish is granted."
"My second wish is not to be a buccaneer anymore."
"Granted", said Neptune solemnly.
"My third wish is to be at home again", I blurted out.

"Your third wish is granted", said Neptune. He waved his trident and I was driven through the waters so fast that I had to close my eyes. When I opened them again I looked into my familiar face in the triangular mirror.
"That was close!" I said to myself.
"What was close?" asked Three who had heard my three words.
"Oh nothing", I replied. "You happen to have a funny mirror in your bathroom."
Soon after that I took my leave. Marching along the Promenade of Numbers, the ground was still swaying for another three minutes. But upon arriving home, everything was all right again.

In Land Three

Three steps are leading right to my door
just knock it three times, that opens the door
I´ll sing with three tones my little song for thee
bring along a friend, we´re singing for three

Garden of Numbers up to 5

The third term has the motto: Who or what may enter Three´s garden? A tricycle yes, a bicycle no. All right are triangles, a picture about triplets, a trefoil leaf, a small pastry fork with three prongs or even a three-master.

A strategic game on Number Lane

This ambitious match of numbers presupposes the comprehension of possibilities with three numbers: 1 – 2 – 3. Yet this shouldn´t be told the children but the game should be repeated until the children find out the strategies for themselves.
Form two groups and determine the length of Number Lane, 10 is advisable but 9 or 12 are also possible. Draw lots for the starting team. Children will act as "living pieces" in the game. One or two pieces of one group may be placed on one respective number, this must always be decided and discussed by the whole group. Winner is the group who places the piece on the last number of the row. Which group may be the winning one depends on two factors: Which group is the first to place one or two pieces on the first number, and which group will develop the best strategy of setting pieces.
Of course it is possible to play this game with ordinary pieces, but children love to act as "living pieces".

Magic Three

The children are milling around in Number Land. The Fairy of Numbers calls aloud: "Three!" or claps her hand three times, or shows, to all clearly visible, the puppet with the number 3. The children must build groups of three and hold hands. Children who remain outside of a group of three may come to the fairy and hold hands. Forgetmenot, of course, is free to clap her hands only twice, or four or five times, but only number 3 does the magic.

Bottlemusic

Three bottles of equal size, filled with the same amount of water are producing, if properly done and blown into, nearly the same sound. Filled differently, the sounds will be different. This, of course, can be varied. There are many possible ways to play the filled bottles:

+One child blows, one after another, the three different tones: high – middle – deep, high – deep – middle.
+The children guess, with closed eyes, the different tones.
+Three children produce with the bottles a triad.
+The children play the melody of song number three.

Telephone operator

You may be lucky enough to find somewhere a telephone with a preselection dial. This dial is suitable for many a different play. Children stand in a row before a desk.
One child is sitting at the other side and plays telephone operator. The first child in the row, opposite of the operator, is asking for three numbers: "1, 3, 5." (Most children can memorize three numbers in a row, more than three may be difficult.) The operator dials the numbers. Then the roles change and the operator becomes subsequently last in the row of waiting children.
Other possibilities with a preselection dial: Dial a number smaller than 4. Dial a number bigger than 6. Dial a number between 1 and 4.

Right through the middle

Three children each form a team. Two of them are standing between 1 and 3 meters apart, the third runs through the middle and stops. Now one of the other two children runs through the new created middle. The children of each team are doing this alternately, this funny game can be played very long.

Activities during the week

+Number Breakfast: This week, for example, each child receives three slices of an orange or apple.
+Number Lane: Number 1 receives one pebble or marble, number 2 receives two, number 3 receives three. This system can be played up to number 10. 55 Objects are necessary for all ten numbers (pebbles, marbles, nuts or such).
+Throw dice one-two-three.
+Book of Numbers: Draw or paint pictures of triple things, for example a tricycle, three feathers, three birds etc.
+Invent and tell new stories.

Four - Four feels ill

I have again taken a stroll on the Promenade of Numbers. I passed the quaint houses of One, Two and Three and finally came to another nice little house. There I stopped and took a look across the fence. And what did I see?
The little house of number Four! But what a sight! It was completely crooked and resembling the hut of Hodgeypodgey. The door was wide open and I could see into the living room. There Four was laying on the sofa. The poor number had a thick bandage around its head and a thermometer in its mouth. I heard it groaning: "Oooh, oooh, that hurts!"
I hurried to number Four worriedly. "What happened to you? Are you ill?"
"Yooo, yooo, yooo, yooo", was its lamenting answer. "I am miserable. Ever since I have returned from my fourth trip. My head is aching, my nose is running and my legs are feeling like blocks of ice. Yooo, yooo, yooo, yooo, poor me!"
I had to smile. "Come now. It can´t be that bad. You´ve just cought a cold. I´ll make you a good tea and everything will be all right again."
Four nodded its head four times and buried its nose into the pillow. I hurried into the kitchen and collected all ingredients needed. Well, at least I tried to do so, because the kitchen was in a total mess. The four-cornered table stood on three legs, one of the four teacups lay broken on the floor and as I put the four-cornered kettle on the stove I just glimpsed Hodgeypodgey leaving the garden over the fence. He munched laughingly at a bar of chocolate, riding his bike and went off.

"No wonder", I thought, "that the kitchen is such a mess. That goblin was here!" Then I went into the garden to gather the herbs needed for the tea. Four leaves of peppermint, four chamomile flowers, four stems of chive and four bulbs of liquorice. Returning to the kitchen I put these ingredients into the boiling water.
Well, children, have you memorized how many ingredients I have collected for the tea?
After the tea was drawing four minutes I sweetened it with four spoons of honey, stirred it four times and said:

Hodgeypodgey, what a dare,
tricks my numbers from here to there,
come to help, Forgetmenot,
chase that rascal from this lot!

As soon as I finished reciting the spell, the boiling tea started shining. And it didn´t smell anymore only of peppermint, chamomile, chives andliquorice, no, it smelled of forgetmenot. "Thank you, dear fairy", I said and served the tea to the poor ill Four.
Four had four sips, got up from the sofa, tore the bandage from its head and shouted: "I´m well again!"
"That´s fine", I said, "where did you catch that cold anyway?"
"I´ll tell you. But come to the kitchen, it´s more comfy there."
Entering the kitchen I was overwhelmed: Everything was in best order and standing in its right place. The table had four legs, the cup was whole and all was well.
"Here we are", said Four cheerfully, "make yourself comfortable."
"I am often away, you must know. To be precise, four times each year."
"And where are you going for your vacation?" I asked.
"I have a nice little holiday home", explained Four, "with a square garden four times by four yards big. And there is always work to be done. I plough it in the springtime and sow four different kinds of seed. In the summer the plants get four liters of water each, for better growing and not to perish of thirst. In the fall I gather fallen leaves blown by the wind in four directions: East, West, North and South. And in the winter I burn the wood in my fireplace against the biting cold or else it would be getting as frosty as in the Antarctic. On my last trip I had forgotten my socks, my hat, my shawl and even my overcoat. That is the reason why I cought such a cold. Fortunately you came by and healed me. For that I will give you a gift."
Four asked me to follow it into the garden. It bent over and was looking for something. Finally it found what it was looking for and handed it to me. It was a four-leaf clover.
"It will bring you luck!" said Four.
I was very pleased with the gift and took my leave from Four shaking its hand four seconds long.

In Land Four

Springtime, summer, fall and winter
are four good times for a sprinter
fire, air and earth and water
do not well without each other,
four corners has my little home
I invite four friends to come!

Number Garden

Who or what may enter Four´s garden and why? A sheet of paper (4 sides), a picture frame, anything square, a picture of a table or chair (4 legs), a toy car (4 wheels), a four-leaf clover etc.

Exercise on Number Lane

With the help of Number Lane children get, step by step, aquainted with numbers. One directed exercise can be as follows:
Children, while walking Number Lane, stop on one number and state aloud which one it is (possibly with eyes closed). This may seem to be a simple task but the children have to interrupt their rhythm of counting and have to start anew with the work of their brain (remembering, counting). This is a serious accomplishment of commemorative work. This exercise should be repeated until the children are mastering it.

The very special die

A die has six sides and therefore shows six numbers. Some games may be played with only the numbers 1 to 4 with two eventualities added: A good and a bad one. These eventualities are represented by the good fairy Forgetmenot and by the goblin Hodgeypodgey. Such a die is easily created: Paste a piece of paper over the sides five and six and paint on them one sad and one happy face.

Gambling

The children form two groups and divide 16 objects (for example marbles, chestnuts, acorns,coins or such) to equal parts. Each group receiving 8 pieces. And now to gambling. Cast lots for the beginning group.The first one throws the special die. As many as the thrown number, as many objects may be taken from the other group. The game lasts until one group loses all 8 items.

Hodgeypodgey throw: 3 items go to the opposing side.
Forgetmenot throw: 3 items received from the opposing side.
Or: Hodgeypodgey:Five times hopping on one leg,
Forgetmenot: 5 times (softly!) tweaking the opponent´s nose.
Or:Let the children find a way to get even with the opponent.

Activities during the week

+Number Breakfast: Each child receives for example four nuts.
+Book of Numbers : Draw and paint a creature with four legs, or of broken things in the house of number Four (who is ill), pictures of the four seasons, etc.
+Discuss the four seasons of the year or the four elements water, earth, fire and air.
+Invent and tell new stories.

Five - Town-Mayor´s Birthday

Yesterday I felt, all of a sudden, a yearning for the Numbers and immediately I took off to visit them. Walking the Promenade of Numbers, something funny caught my attention. All trees, left and right, were decorated. Do you have an idea what that was for?
I passed the gate of the town, came to house number One and knocked on the door. No answer. I went to number Two and knocked twice. Two was also absent. Number Three got three hammerings on its door. I waited three minutes, but again nothing happened. I began to get anxious. But at number Four I was lucky. Immediately it came running.
"I have no time", called Four, "I must hurry. Today is our Town-Mayor´ s birthday and it is planning to celebrate in great style. Come along also, the others are already there!"
I followed Four running. Upon arriving on the marketplace we were both breathless. Everything was decorated and confetti was pouring down.
"Five cheers to our Mayor!" called One, Two, Three and Four loudly and I joined them. But I hadn´t seen Five yet.
"Where is Mayor Five?" I asked Two who was waving two pennants.
"At at home home of of course course", said Two and kept the pennants moving, "in in that that big big house house."
Finally I saw it. It looked a bit uncommon with its pentagonal door and windows.
"Why don´t you go in?" I asked One, Two, Three and Four, "is it forbidden to congratulate the Mayor personally?"
"Quite the contrary", said Three and threw three confetti in the air. "We are all invited but we can´t go in!"
"This I don´t understand", I said. "You are throwing confetti, waving pennants yet can´t enter the house?"
"Exactly", said One, "although we are trying everything. Maybe Five will see us through the windows and will join us out here."
Then somebody poked me from behind. "Make room!" called a strange boy.
"Make room yourself", I retorted, "what impoliteness!"
"Please excuse us", said the boy now politely, "but we are late. Would you please let us through?"
I stepped aside and saw four more children. In addition to the boy, there were one smaller and two bigger girls and an even smaller boy. They passed us and entered Town Hall by the pentagonal door.
"That´s mean", protested One, "we also want to go in.
Wait for us!" But the five children were gone.
"What shall we do?" asked Three annoyed.
"I I don´t don´t know know either either", said Two enraged.
"But I do", I said and smiled. "I know how you´ll be able to enter."
"Tell us!" called One, Two, Three and Four expectantly.
"Very simple. You must try to make five."
"How can that be done?" asked One.
And that was a difficult task to be solved by the four numbers.
Could you help them?

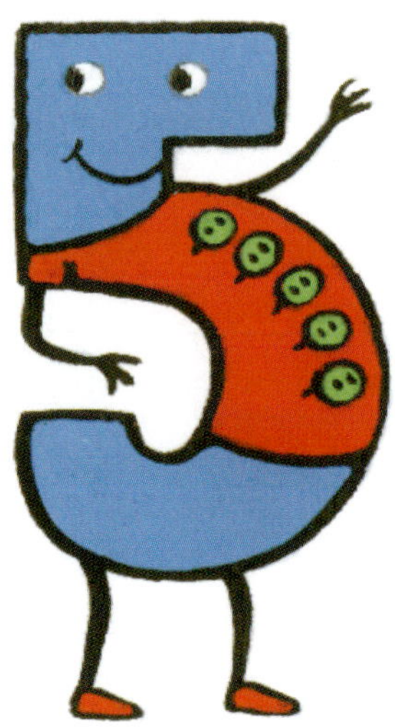

Right! One took Four´s hand, because one and four are five. Two took Three´s hand, because two and three are also five. And this is the way they passed the pentagonal door. I gathered five confetti and followed them. Finally all guests were together.
Five was sitting at the upper end of a long dinnertable. Its guests just stood around somewhat embarrassed. Finally One plucked up courage, went to Five and gave it a drum as a present.
"Happy birthday", said the little number with a smile.
Then came Two with two small silver bells, Three with three colorful feathers and finally came Four with its present: an apple, a pear, a carrot and a walnut. "All were grown in my garden", said Four.
Five thanked the numbers for their presents and then introduced the children: The first boy came from Europe. He presented Five a blue flag with a circle of golden stars. The small girl came from Asia. Her present consisted of five pairs of colorful chopsticks. One of the bigger girls came from Africa. Her present were five brown coconuts. The other boy came from Australia. He presented Five a wooden kangaroo. The last girl came from America. Her present was a hat adorned with five beautiful eagle feathers.
Five rose to its feet and said: "You came visiting me from five continents and my friends One, Two, Three and Four are here also. Let us be merry together!"
And we celebrated till five o´clock in the morning!

In Land Five

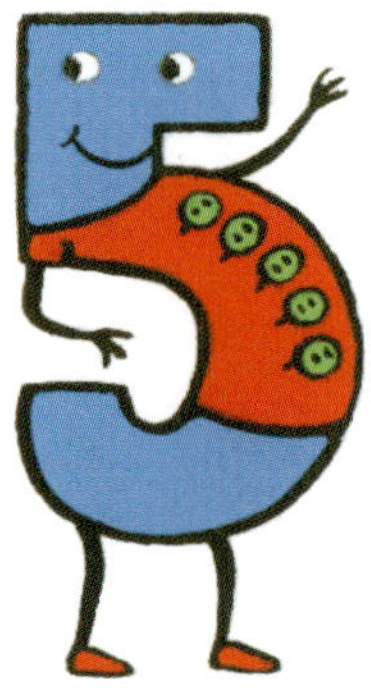

Five corners has my house
in it lives a little mouse
five fingers make one hand
I live in Number Land
Five are more than Three
please come and visit me!

Number Garden

Who or what may enter Five´s garden? Gloves with 5 fingers, a pentagram and of course, a starfish. Who finds a flower with 5 petals?

Exercise on Number Lane

If possible, don´t raise the degree of difficulties before the ones of number four are completely understood by the children. A possible next step is to let the children walk Number Lane blindfolded. When standing, they should say what number they are standing on and which ones are before and behind them.
An idea: Children are losing their pleasure to play very fast if they are continually advised. Use instead the principle "learnig while doing": Walking Number Lane and counting aloud: "1 – 2 – 3 .... I´m standing on 5, what is behind me, what in front of me? Who can help me?"

"Attendant"

The entrance to Five´s garden is supervised by an "attendant". Only items or persons with the number 5 may enter. The attendant is asking for all that is 5 on the entering person: 5 fingers, 5 toes, 5 points glued on a nose etc. Or 5 children holding hands may enter. Or 2 plus 3, 4 plus 1, etc.
An idea: A real entrance may be constructed by glueing together cardboards, thus forming an "Entrance".

Find out by feeling

Pouches can be filled with different items, like hazelnuts, acorns, marbles etc., children are then asked to put their hand into the pouch and take out a given number of items, 1 ,3, 5, etc. Or place the pouches on the numbers in Number´s Row and compare their contents with the number chosen. Another possibility is to fill the pouches with self-fabricated geometrical forms. Circles, squares, triangles etc., made of cardboard. Children have to feel with their hand and find out what form they are grabbing.
Feeling with bare feet also is possible, children love to examinate things with their feet. Blindfolded they could find out with their feet, for example, how many marbles or other objects are hidden in a plastic bowl.

Activities during the week

+Number Breakfast: Each child gets, for example, five radishes.
+Book of Numbers: Draw or paint a starfish.
+Look at the 5 continents on a globe.
+Children on these continents – how do they live, what do they eat and play etc.
+Thematize the five senses.
+Strings of Numbers: These can be made with pearls of glass, different colors, chestnuts, acorns. Children love to make these strings.

Six - In the Desert of Forgetfulness

The other day I visited Number Land again. I had barely arrived, when I saw Hodgeypodgey approaching on his bike. "What do you want?" I asked him.
"Oh, nothing, really nothing", he said and sped off on his bike laughing loudly.
That gave me a funny, foreboding feeling. Lost in my thoughts I followed the Promenade of Numbers, made three steps forward, four steps backwards, one forward, then again five backwards, two steps forward, two again backwards and – oh my goodness! – all of a sudden I was lost. I went completely astray. I looked around and everything was gone. The Promenade of Numbers, the Town´s Gate and the whole town. All these had vanished, simply, totally disappeared. I was surrounded by sand dunes and the sun was scorching my face. I had stepped into the Desert of Forgetfulness! There were no trees and no shrubs to be seen! Nowhere was there a shadow. The awful heat made me terribly thirsty. In six minutes my mouth was completely dry. What could I do to escape this desert?
Would you, children, have an idea? Come, let us all together recite our spell:

Hodgeypodgey, what a dare,
tricks my numbers from here to there,
come to help, Forgetmenot,
chase that rascal from this lot!

All of a sudden I heard somebody sighing. It sounded so sad and so weird. It didn´t sound like the good fairy Forgetmenot at all! Hurriedly I climbed a sand dune. And what do you think I discovered?
A number was standing there! In front of it stood an easel with canvas. In its left hand it held a palette and in its right hand a paintbrush. Every once in a while it painted, sighing, a colorful spot on the canvas.
"Hello!" I called to the number, relieved to see somebody.
The number turned around. "Hello", it answered, "I am Six, and who are you?"
"I am ...., I got lost!" was my reply. "Besides I am terribly thirsty."
"Come over here", beckoned Six, " I have something to quench your thirst."
"What a lucky thing! Had I not met you, I´d have perished from thirst!"
"You are indeed lucky", said Six and handed me a big bottle full of water. "Were it not raining in Numbertown, I wouldn´t be here."
"I don´t get it", I said after having taken six big gulps of water. "What has raining in Numbertown to do with your being here?"
"Just have a look over there", pointed Six, "and you will understand."
I looked into the direction Six was pointing. What was that? Heavy dark rain clouds were hanging over the horizon. Above us bright sunshine and there rain clouds! And in front of those heavy clouds stood a colorful rainbow, painted there by the rays of the sun.
"Fantastic!" I called. "Now I understand why you are here. You want to paint a picture with a rainbow in it!"

"That is right", said Six, but it sounded somewhat gloomy. "Unfortunately I do have only three colors, red, blue and yellow. How can I create a rainbow with them?"
"Hey, I know a way to help you!" I called. "Just give me your palette and the brush."
On the palette I found the colors red, blue and yellow. First I cleaned the brush with water, then I painted a red, a yellow and finally a blue spot onto the canvas. Now I had a triangle of three colours near each other on the canvas. I took some water with the brush and mixed red and yellow together.
"Wow!" called Six astonished, "mixing red and yellow makes orange!"
I nodded and mixed red and blue.
"Mixing red and blue gives violet!" Six was surprised.
Finally, when I mixed blue and yellow, Six shouted excitedly.
"That makes green! Yippee! Red, orange, yellow, green, blue and violet! Now I can create a rainbow."
After the painting was finished we wandered towards the rainbow and Numbertown. You can´t imagine, how happy I was to be able to leave the Desert of Forgetfulness! Together with Six we practically danced in the rain, jumped into the puddles and admired the rainbow which shone above the hexagonal house of Six. Its six iridescent colors red, orange, yellow, green, blue and violet had luckily led us back from the Desert of Forgetfulness.

In Land Six

Six steps will suffice to bring you to me
six tones has my little song for thee
a rainbow stands over my house in the sky
six colors shining in the heavens up very high

Number Garden

Who or what may enter Six´s garden? A die (with 6 surfaces), drawings of a bug (6 legs), honeycomb (6 corners), drawing of a rainbow (6 colors) etc.

Exercise on Number Lane

Arrange the numbers at random. Children search for the right sequence and jump from number to number, starting with 1...

Win and take

A certain ammount of objects (marbels, coins, nuts) are spread on a tabletop. Each child in succession throws the die and takes as many objects as the die shows. Is the last object taken, children count their spoil. He/she who has most, has won.
This game can be reversed: Each child receives a certain number of objects (10, 12 or 14) and lays them aside according to the thrown number. The first child without an object is winner. A special degree of difficulty: The last throw must match the number of the remaining objects.

Who is first?

Many games of dice may be played after knowing the number six. Example: Each child has six cards laying in front of her/him, with the numbers 1-6 depicted. One child throws the die and the others hold up the card with the thrown number.

Submarine

This amusing game needs a die, a small basin, a spoon and something that can swim (top of a small glass container for piccles, or such). The "submarine" is placed on the water. The children subsequently throw the die. The number thrown means the number of spoonfuls of water put into the "submarine". The child who sinks the "submarine" is winner.

Rainbow roundelay

The children fashion, using fabric or crepe - paper, six or twelve ribbons in rainbow colors and some in white and black, these latter representing sunshine and rain clouds.
We are playing weather: Each child receives one ribbon and whirls around in the room or outside. On a given sign six children unite to a rainbow with their colors (ribbons). The remaining children stand in a group left and right, representing sunshine and rain. Then the ribbons (colors) are changed so that each child represents, in turn, rainbow and weather.

Activities during the week

+Number Breakfast: Each child receives for example six slices of pickles.
+Book of Numbers: Draw or paint a bug with six legs.
+Learn the names of different bugs (May beetle, ladybird, potato beetle etc.)

Seven - Disorder in Fairyland

First of all you must know that numbers know the same fairy tales as we do: For example, think of The Wolf and the Seven Little Kids, Brave Little Tailor and Snow White. Do you know how many dwarves were with Snow White?
I had planned to visit Seven and bring her the fairy tale about Snow White and the seven Dwarfs. But as I looked for my storybook it was gone! I searched the whole house over, but couldn´t find it. And so I had to leave for Number Land without my storybook, because I had an appointment with Seven, on the seventh day of the week at seven after seven. Do you know on what day I had my appointment with Seven? (On Sunday!)
And so on Sunday the seventh day of the seventh month of the year at seven o´clock I rang the seven small bells of house number seven in Number Land. Nothing moved. I was quite disappointed because I had made haste to be in time.
But listen! Did I hear the bell of a bike? Indeed! It was Hodgeypodgey driving by on his rusty vehicle. He grinned, threw a small package over the fence and sped away. I picked the package up. A scrawly seven was written on it. As I wanted to put the package into Seven´s mailbox, many voices began singing loudly: "Hay-ho, hay-ho, we are the seven dwarfs and sing hey-ho, hay-ho!"
But where did the singing come from? And again I heard: "Who hid the two saucers away? Where are the four small beds?Where have you laid the little forks?"

There was something fishy about these things. Should it not be seven saucers, seven small beds and seven little forks? And other voices chimed in. That of the Wolf and the Seven Little Kids, with only five of them mentioned, and Brave Little Tailor who in this version had caught only five flies with one stroke. And then it dawned on me: The voices came from the package Hodgeypodgey had thrown over the fence!
Hastily I opened it and guess what I found inside? My missing storybook! Hodgeypodgey had sneaked into my house and had clandestinely taken it away. Such a naughty goblin!
"Help!" called a voice suddenly from inside the book. "Could somebody help us, please?"
I opened the book immediately and discovered that it was in a terrible muddle. The dwarfs, the kids and the flies ran around headlessly.
"This is terrible", I said, "Who could and would help?"
"The right number", whispered the voice of Snow White. But where could that be?
Desperately I recited the spell:

Hodgeypodgey, what a dare,
tricks my numbers from here to there,
come to help, Forgetmenot,
chase that rascal from this lot!

Everything lit up at once and I had to close my eyes. When I opened them again, Forgetmenot stood there.
"Help me please, dear fairy", I begged in despair. "Hodgeypodgey has mixed up all the tales and Seven is also missing."
Fairy Forgetmenot smiled and said: "Don´t you worry." She opened the door with a wink of her wand and called loud: "Good morning you sleepyhead!" and disappeared in a flash.
"Who is disturbing my sleep?" said a grudgy voice. Seven, clad in a dressing gown, came shuffling out of the house. Upon seeing me Seven threw up its hands in astonishment and called: "Is it already seven o´clock?"
I just nodded.
"I´m sorry", Seven said bashfully. "I have overslept our appointment!"
"I must also excuse myself", replied I. "Hodgeypodgey pilfered my book and the tales are now all in a disarray!"
Seven looked into the book and said: "That rascal Hodgeypodgey has tricked us both. He charmed me out of every tale while I was asleep!" Seven shrank, became very thin and slipped into the book. It brought the dwarfs seven saucers, seven small beds and seven small forks. It found all seven kids and caught all seven flies of the Brave Little Tailor. Seven established order again with the tales and promised to set its clock in the future and not to oversleep.
We spent the remaining time of the day in Seven´s garden and ate seven dwarf cakes that Forgetmenot gave us as a present.

In Land Seven

Seven steps you have to go
to reach my house and say hello
seven chimes are at my door
seven flowers come before
seven days go hand in hand
that is one whole week here in Number Land

Number Garden

Who or what may enter Seven´s garden? One side of the weekly calendar (7 days), a ladybug (most ladybugs have 7 points), a starflower (7 petals) etc.

Exercise on Number Lane

Prepare Number Lane up to numbers 10, 15 or even 20. Turn on the floor as many numbers as children are present upside down. Each child must guess one hidden number it is standing on. The best strategy is for the children to walk the row and count aloud or silently. This task is, as of this time, for most children soluble and joyful.

Which number is concealed?

The children are sitting around a table and each child in turn is throwing the die. Then it calls the number it has thrown. This number of course is laying on the upper side if the die and the answer is easy. But who knows the opposite, the concealed number?
The answer of course is not an easy one, especially for children not used to play dice. (The solution: The sum of the open and the concealed number is always 7). Have children once found out the "secret", they most likely will enjoy this game of finding out the concealed number.
Variety of the game: Children are sitting in one row and are looking at the die from a different perspective. The question now is: Which number is concealed and which one is opposite of the one they are facing?

One week game

Two groups of 7 children each, with signs on their breasts, each sign naming one day of the week. The task for the children is to line up in the sequence of the week (Monday, Tuesday etc.). Another possibility: The name of the day is being whispered to the children and in turn the children whisper the name of the day and find the right arrangement. Which group is first and the winner?

Blind frog catches fly

As first step determine, by a counting rhyme, who is the frog. The frog is then blindfolded. He must catch his fellow players jumping like a frog. This is likewise done by the "flies". Thus they can be heard and located by the "frog". The "flies" may, contrary to the "frog", jump only seven times. This finished, the "frog" may catch a fly more easily. Then the catch turns into the "frog" and vice versa.

Catch the hat

Before the game starts each child is named after a number, 1 to 7, or above. Each number is doubly represented (Ideal is a group of 14 children for this game). The children best display their number on a headband or a breastplate. Next step is to determine the catcher, any given number. The catcher puts a hat on and wanders around the room. Then she/he calls out: "Five has lost the hat and number two (or whatever number) is getting it!" Quickly one of the children with the called number must get the hat and is continuing the play. ("Two has lost the hat and ... gets it!")

Activities during the week

+Number Breakfast: Each child may receive for example 4 nuts and 3 radishes.
+Book of Numbers: Draw or paint a picture of something having 7 limbs / points / corners etc.
+In many German tales the number 7 plays a significant role as do other numbers in other cultures and tales. It can be a nice idea to go on a number hunt in tales or stories.
+Tell a fairytale: Snow White and the 7 Dwarfs.
+Learn the seven days of the week.

Eight - At the Fairgrounds

It was a sunny day in beautiful Number Land. Eight stood in front of its little house, stretched its arms in the air and said: "I´ve worked now for seven days but today I want to relax and do something crazy."
Immediately it collected its eight things: a backpack, a purse containing eight Euros, a bottle of water, a snack, an umbrella, a handkerchief, a cough drop and a bus ticket. Then Eight took off. It wanted to go somewhere, where it, and we also, could have much fun. Do you have an idea where Eight planned to go?
Eight planned to go to the fairgrounds! Wow, the things that could be seen and found there! The first thing Eight discovered was the stand for cotton candy, second the merry-go-round and finally a Clown selling balloons. Further in the back a long row of people was waiting for something.
Eight asked the Clown: "What kind of a stand is that over there?"
The Clown immediately recognized Eight. "Oh, hello Eight! Don´t you know what that is? That is the main attraction: The famous, phantastic roller coaster!"
"What is it for?" asked Eight.
"You will see", answered the Clown. "Just take a seat in one of those carts."

Which Eight did and off it took like a lightning. First came a left turn, then came a right turn, then up-up-up it went, down it scooted. And up again, but this time head down. Wow! Eight minutes later the vehicle came to a stop. Eight tried to get up and leave it but things were turning before its eyes and it got white in the face. Luckily the Clown came around and helped it to leave the roller coaster.
“Do I ever feel funny”, said Eight.
The Clown laughed and said: “Everyone gets dizzy after a ride in the coaster. Especially after it was looping the loop like a number eight!”
But soon the color had returned to Eight´s face. “That was really fun!” it said happily.
“There is more over there”, pointed the Clown. “Come to the one where you can see how strong you are. The game is called Punch the Luke in Germany, where I am from.”
“What?” called Eight indignantly. “I should punch somebody? Never in your life! First, I don´t know anybody named Luke, second, he has never done anything to me, third, it is quite dumb punching somebody, fourth, I don´t feel like hurting somebody, fifth, who is this Luke anyway?”
“Sixth”, continued the Clown laughing, “it is much too hot, seventh, I make you a present, a balloon, and finally, eighth, come and see. You´ll have plenty of fun.”
“All right”, agreed Eight, “but I won´t punch anybody!”
"Come closer! Who will hit the Luke?"
Eight became curious and followed the voice.
“Hello, whom do we have here?” said the barker upon discovering the Clown and Eight. “A balloon vendor and a, ... you are not a human being, but what are you then? Two circles, one on top of the other, meeting in the middle. But I do know you from somewhere. Turn around for a better inspection.”
“That´s going too far”, thought Eight and said: “Even if I´d be standing on my head, I´d be Eight!”
“No offence meant!” replied the barker, “Luke is over there. Please come, gentlemen.”
Eight and the Clown followed the man into a tent, in the center of which stood a wooden block with a rod and the numbers one to eight painted on it.
The barker handed Eight a heavy hammer and encouraged it: “Give it a blow!” and pointed at the wooden block.
”Oh, this is Luke”, said Eight laughing, raised the hammer and brought it heavily down on the block. “Bang!” A pointer buzzed up the rod and hit the number eight.
“My dear fellow!” the barker called astonished. ”One stroke and you hit first prize!” He shook Eight´s hand approvingly and handed it a carillon with eight small bells. "See, the game is also called Ring-the Bells and now you have a carillon with eight bells on it."
“Thanks a lot”, said Eight pleased and let the eight bells chime. “Thanks, but I must go now. It has become late.”
The Clown looked sad upon hearing this.
“I´ll come again for sure”, said Eight to him, “and then I´ll bring all my friends from Numberland with me!”
The Clown looked happy again. “The roller coaster will be waiting for you”, he said.
“And so will Luke!” completed the barker. “See you!”
Completely satisfied, Eight turned to go and smiled happily. It thought: “What a story I have to tell. The other numbers will be so surprised!”

In Land Eight

1 2 3 4 5 6 7 8
you don´t have very long to wait
just step in and sit right down,
Eight is my name and I am a clown
of two circles I am made
turn me upside-down
I´ll remain the Eight!

Number Garden

Who or what may enter Eight´s garden? A die (eight corners), a roller coaster (looks like an 8), and all spiders, for they have 8 legs.

Number Gardens

Exercise on Number Lane

It is by now definite that Number Lane is an excellent means of experience for kineastic operations of counting by moving. That is, moving forewards is adding, moving backwards is subtracting. The order "plus 2" means to advance 2 steps, and "minus 3" to go 3 steps in the other direction. It is advisable though to rather say foreward or backward instead of "plus" and "minus". A play of counting can begin, after the children have mentally connected "addition" and "subtraction" with moving forewards and backwards on Number Lane.
All children stand in a row before Number Lane. The first child throws the die (a possibly large one), hops to the number it has thrown and places something (marble, stone etc.) on it. The next child starts where the first one has placed its object, and throws the die. An adult (or even the Fairy of Numbers) is directing the play (Foreward or plus, backward or minus). Thus the play is always in movement. An arranged number may be set as final target. But children will like this play anyways.

Come to Number Land - game

Waterglass Music

Eight waterglasses, each to be filled with so much water as the scale (sol-fa) requires. The less water is in the glass, the deeper the tone. Wet the rim of the glass and produce the 8 notes. If this is too difficult, you may try producing the notes, very carefully, with a small teaspoon. Each child produces tones with the help of a glass. Key in major produced by 8 children. Many chords and intervals can be produced. The listeners must find out the numbers of tones produced. All 8 songs of numbers can be played.

Activities during the week

+Number Breakfast: Each child may receive, for example, 4 slices of a mandarin and 4 grapes .

+Book of Numbers: Draw a spider with 8 legs or a roller coaster with an 8.

+Tug-of-war: Two teams with 4 children each play against each other.

Nine - Crash landing in the Forest of Errors

Do you know that Earth is a round planet and rotates around the sun? It needs one whole year for one orbit! You may understand the idea perhaps easier if you imagine a tiny bee flying around a huge sunflower. The bee is our earth and the sunflower is the sun. And there are many other insects flying around the sunflower: butterflies, bumblebees, moths and winged ants. Our sun has nine planets rotating around her, like tiny insects. One of the planets is ours and is called the blue planet Earth.
The other day, while I was leisurely walking along the Promenade of Numbers in beautiful Number Land, a shadow fell on me. Astonished I looked up into the skies. A beautyful colorful captive balloon was floating nine meters above my head!
Have you ever seen a captive balloon?
It was held together by a net of ropes which also were holding a basket. Somebody was leaning over the side of the basket and waving to me.
"Hello, you down there, could you please get hold of the rope of my balloon?"
"Of course", I replied and tried to catch the rope. But it eluded me and swung away in front of my nose. The wind had gotten hold of the balloon and blew it further away. Every time when I tried to grab the rope, it moved away and I had to run after it. I did this over and over for nine minutes.
"I can´t make it!" I cried breathlessly.
"But you must!" cried Nine back. "Or else I´ll have a crash landing in the Forest of Errors!"
"I´ll try to get some help", I wheezed back to Nine and ran off to Number Town.
"Please hurry up!" begged Nine.

Upon arriving at the marketplace I started hollering: "Hello, all numbers! Come to me! I have a case of emergency!"
Surprised, the numbers came out of their houses.
"What what happened happened?" asked Two frightened.
"How can we help you?" asked Eight.
And Five said: "Emergencies are my job. Come to Town Hall and make a report."
"Unfortunately there is no time for that", I replied hastily. After having collected One, Two, Three, Four, Five, Six, Seven and Eight around me, I told them the story. "All of you must help, or else Nine will be crash landing in the Forest of Errors", I urged.
"Oh my goodness!" cried Four. "Nine has returned from its trip around the world and can not land without our help! Come on all and hurry up!"
We saw, from afar, how the wind was blowing the balloon towards the trees.
"Help!" cried Nine desperately.
We ran as fast as we could, trying to get hold of the rope. One and Eight got hold of it first. But all of a sudden they were swept away. Two and Seven were next to grab the rope but even they could not stop the balloon. Three and Six gave all their strength, but to no avail. They were also dragged along. Finally Four and Five caught the rope. All the numbers together successfully kept the balloon from flying further. With their combined efforts they were able to tie the rope to the first tree of the Forest of Errors and Nine could, luckily and happy, disembark.

"My, that was close!" Nine swept the perspiration from its forehead.
"You may say that again!" called Five.
"Where have you been so long?" asked One.
"I´ve orbited the world", reported Nine. "Just like the nine planets orbiting the sun, I´ve done the same in nine months with our Earth."
The numbers didn´t quite understand.
"What´s that – planets?" inquired Seven.
"Planets are huge globes, resembling our Earth. They sail through the universe and rotate around the sun. Got it now?"
"Not quite yet", admitted Seven.
"Let´s play planets", proposed Nine. "Then you will see what I mean. For that purpose you must form a circle. You", and Nine pointed at me, "are the sun. Stand in the center."
I obeyed and placed myself in the center of the numbers who were forming a circle.
"And now", said Nine laughing aloud, "we are going to dance around you. Each one of us is a planet!"
Oh my, was that ever a jolly dance of the planets! Nine numbers rotating around me! It made me quite dizzy. Following that, Nine invited us to a balloon trip, but we all declined. Who knows? At the end we may have all landed in the Forest of Errors!

In Land Nine

Nine round planets are circling
around the smiling sun,
nine merry children are circling
around and having fun,
in lovely Number Land
It takes nine
steps to me on the floor,
if you want to visit me today
just knock nine times on the door!

Number Garden

Who or what may enter Nine´s garden? A tower built of nine empty matchboxes a field of nine drawn numbers, anything else pertaining to the number nine.

Balloontrip round the world

All children travel with the kite balloon around the world. All stand in a circle, holding hands, and - whooosh - , off we take, flying over the countries of the earth, landing every once in a while in 9 different countries. What adventures do we have in those countries? What do people look like there? How do they live? What are they eating? What kinds of games do the children play in them? All children deliberate, look at pictures etc. Then take off again to finally land in Land Nine, Nine´s garden.

Bowling

All games of bowling have but one aim, to bowl over with one throw as many pins as possible. You can make pins out of plastic bottles filled with water or sand, to play outside. Bowling is always played with nine pins. Her or him who bowls over the most pins, is winner. Variation: You can mark the pins with numbers or even names and winner is whose pin is surviving the throws.

Number Land - Domino

Hopscotch

Hopscotch comes in many different versions, many of them forgotten but worth being rediscovered:
The numbers One to Ten have to be reached and touched by hopping and jumping. The numbered fields may be marked by marbles or small stones for better remembering. Please find new variations!

Activities during the week

+Number Breakfast: Each child receives, for example, 5 slices of an apple or a banana and 4 nuts.
+Book of Numbers: Draw or paint, for example, nine pins.
+Learn the planets of our solar system.
+Listen to the Symphony Number Nine by Ludwig van Beethoven

Ten – In the Labyrinth of Stairs

The other day I was again walking along the Promenade of Numbers and counting my steps. 1, 2, 3, 4, 5, 6, 7, 8, 9. By this time I had almost reached the end of Number Town. But look! What was that? A huge decagon of a building was looming ahead. A building of many floors, resembling a knight´s castle stood there. It was surrounded by a moat, with a narrow drawbridge leading to the gate. And who would be living in this tower? Yes, Ten! I decided to visit number Ten immediately. Carefully I crossed the draw-bridge, went to the gate and pulled the doorbell ten times. A window was opened very high up on the tenth floor and number Ten looked down.
"Who is there?" it called.
"It ´s me, ...", I answered.
"My goodness! But that´s very nice of you to come and visit me", Ten called back. "The others have already told me about you, come on up and have a look around. The view from here is just marvelous!"
The gate opened soundlessly, just like magic, and it closed immediately after my entrance. Turning around, I discovered Hodgeypodgey behind heavy bars. He snickered: "Now you´ve been caught! You´ll never escape from here."
"Dont´talk nonsense", I replied. But secretly I felt uneasy. What had he meant by *I would never escape from here*?
A steep staircase led to the upper floor. I went up and came to another door which I passed. It led to another staircase going left and right. Which one should I take? I decided for the left one and had to pass another door. Here I came to staircases leading into three directions. I took the one in the middle and again had to pass a door. "Oh my goodness!" I said astonished, because I was standing at the foot of the double staircase where I had started my climb.

Behind the bars Hodgeypodgey was snickering: "I´ve told you so! You´ll never escape from here, from this labyrinth of stairs."
Do you know what a labyrinth is? A labyrinth is a great big number of ways and roads leading crisscross from one place to another. Some of them end abruptly, facing a wall, others just go in circles round and round. Actually there is only one way leading out of the labyrinth. And I was caught in such a labyrinth. What do you think I did to escape this? I recited the well-known spell:

Hodgeypodgey, what a dare,
tricks my numbers from here to there,
come to help, Forgetmenot,
chase that rascal from this lot!

And Forgetmenot appeared at once.
"Dear fairy", I pleaded, "please help. Hodgeypodgey has locked me in!"
Forgetmenot handed me a ball of red thread.
"What am I supposed to do with this?" I asked her.
"You´ll find a way. Just think", she said smiling and - whoosh - disappeared.
There I stood with a ball of red thread and thought things over. After ten minutes I finally had it: I just had to count! The first stairs led me to a double stair, those to a triple one.
I only had to follow the number which led to the next stair. Forgetmenot had given me the ball of thread to mark my way! I knotted my thread to staircase number one, at the bottom, and began my climb upwards. Now I could not miss my way. After a while I reached the tenth floor.
"Here you are and welcome!" called Ten upon seeing me.
"Whew!" said I. "So many stairs makes one breathless and confused."
"But in spite of all, you´ve made it", Ten praised me.
I stepped into a room with ten corners and ten balconies around. The view was really breathtaking! You could see all of Number Land, the Sea of Infinity, the Promenade of Numbers, the Desert of Forgetfulness, the Forest of Errors and Number Town.
Looking at my watch I saw it was ten after ten. "I´m sorry", I said to Ten, "but it is late and I have to go!" and hurriedly descended the steps. With the red thread tied to them I could not go wrong and arrived happily outside.
Hodgeypodgey was standing there and upon seeing me he looked disappointed, stuck his tongue out at me and rode away, in the direction of the Forest of Errors. Ten waved and called from the tenth window good-bye and I waved and called back. Then I hurried home.
This was my last story about beautiful Number Land. Now it is up to you, to visit the numbers. I am certain that you shall have many exciting adventures with them! And maybe you will tell me about those adventures some time.

In Land Ten

1, 2, 3, 4, 5, 6, 7, 8, 9, 10
big steps have you to make to reach my den,
steep is the stairway and the road is long,
just ten tones and finished is my song!

Number Garden

Who or what may enter Ten´s garden?
A tower built of 10 empty matchboxes,
a pair of gloves with 10 fingers etc.

Number Land´s Treasure

Ever since the numbers can think, there is a treasure hidden in Number Land! The Fairy of Numbers, Forgetmenot, has hidden it well in a treasure vault. Every once in a while Hodgeypodgey, the Goblin of Numbers, gives it a try to get to the treasure, but for the Magic Power of Forgetmenot. Except for this time. This time he has succeeded. All members of Number Land are greatly excited, because to them this is a great loss. But Hodgeypodgey, while fleeing Number Land, has lost the treasure. Fortunately, though, you and your friend have found the treasure. Yet, instead of fighting for it, you have decided, being very fair, to have a competition game. Whoever wins the treasure may return it to the treasure vault. Then peace returns to Number Land.

Rule of the game

First: The story is being read aloud. Then the treasure (a cassette, a pearl or nice stone etc.) is placed on the starting position. This treasure is used as the piece. A lot is cast for the beginner. Each player choses one direction (left or right in the circle of numbers), throws the die, counts in the chosen direction and moves the treasure-piece to its new place. Thus the treasure is always in movement. Who manages to bring the treasure to the treasure vault?

What do children learn in the Treasure Game?

At this game children don´t have to be able to read the written numbers, but only the number of points on a die. And they have to be able to count their steps up to 10. Learning to read and memorize the written numbers comes in this play additionally. Children playfully learn the arrangement of the first ten numbers as well as mathematical operations, like addition and subtraction to the number 6, or even to 10. After having played this game for several times, children will possibly be able to move the treasure without counting step-by-step.

Ten celebrates its Birthday and receives gifts:

From One: One game of bowling (1 + 9 pins).
From Two: The drawing of a spider (2 + 8 legs).
From Three: A calendar of weeks (3 + 7 days).
From Four: A die (4 + 6 sides).
From Five: A seastar (5 + 5 arms of the seastar).
From Six: A toy car (6 + 4 wheels).
From Seven: A tricycle (7 + three wheels).
From Eight: Two glass balls (8 + 2 balls).
From Nine: One magic wand (9 + 1 wand).
What else could be presented as gifts?

Activities during the week

+Number Breakfast: Each child receives, for example, 5 walnuts and 5 strawberries.
+Book of Numbers: Draw and paint beautiful Number Land.
+Discuss decathletes.
+Invite and tell new adventures in Number Land.

The Promenade of Numbers

A colorful excursion into learning the numbers from one to ten

Children are learning while playing, and they do this with all their senses: they touch objects, examine their colors and tastes, they listen to foreign as well as to well-known sounds and imitate them. They transfer music into sound and movement and put all-day happenings into scenic pictures, thus processing and shaping them for their better comprehension.
Infantile learning is always both integral and playful. Therefore the Promenade of Numbers is leading, together with song and music accompaniment, to an easier learning of the elementary numbers 1 – 10.
In merry rhymes numbers are even talking. They are turned into real persons which makes it easier for the children to remember and memorize them. Within the songs the numbers are connected to daily events in nature and surroundings, thus enabling the children to connect the other elements of different fields of knowledge.
But above all, the Promenade of Numbers – connected to mathematical correlations – provides an insight into the colorful world of music: Each number is associated with a tone of our major key and has a special sound, thus becoming an impressible audio experience. Double and triple rhythms are common in music. They represent the smallest rhythmical units and may be perceived by plays of movements or by clapping hands. Combining these smallest units, other rhythms and numbers may be discovered. The constantly repeated refrain gives assuredness and leads to the next number.
Music and movement, combined with elements of play and learning, provide countless possibilities of presentation. This embracing approach supports creativity in dealing with our system of numbers and eases the children's approach to mathematics.

Structure

The quite different songs to each number are imbedded into a frame story (refrain). The visits to Number Land are all joint efforts. Step is followed by step up to the last number. A finishing refrain allows to leave the world of numbers again and to return to it another time. The songs can be rehearsed and at the end of a given period presented as a whole performance. Some children, depending on individual music talent, may find one or another song difficult to memorize. But the refrain is always offering a possibility to enter familiar fields and from there it is easy to discover new musical structures.

Music Style

Educational theories of music have shown that classical music is able to considerably increase the receptiveness of children. Therefore classical style has been chosen for the works on the Promenade of Numbers. The main tasks are being taken over by instruments like violin, oboe, harp, flute and piano. The songs are rather romantic than modern. Instruments from the realm of Rock and Pop, like drums, percussion, synthesizer and electric bass, are serving as an addendum of Swing.

Lyrics

The short, rhymed poems combine significant and merry ideas and do not need their melodies. Based upon worldwide well-known verses for children, they are catchy and may be considered helpful with merry games. But they create, in all cases, combined with music and singing, deep rooted emotional contacts to perceptiveness and logical thinking of children in their process of learning.

Singing with children

Singing is good for mind, body and soul. Yet music teachers often criticize that music and singing is so rarely taught to children - and this in spite of the well-known fact that, besides gestures and facial expressions, voice is the most direct means of expressing our thoughts and feelings. It should therefore be trained and supported from the very beginning.
Singing loosens our tensions, frees suppressed feelings, creates self-confidence, makes us heard, strengthens our respiratory system, helps us to become aware of our physical self and is beneficial to the sense of being part of a community of musicians.
It is often being discussed which tone level should be used when singing with children. Some children reach very deep, others very high tones, some children squeal, others drone - but all are happy to manifest their voices and all have a lot of fun. Our experience has shown that the best and most "healthy" singing with children is between d´ and d´´. Teachers not being able to follow children to these heights should switch over to an accompanying instrument, instead of singing the songs in a deeper tone and thus probably overburden the young voices.
Sound volume is another important point: Some children enjoy singing very loudly, others very faintly, almost inaudibly. Recommendable is a safe middle way, i.e. neither too loud nor too soft. In any case remember not to expect too much from a young voice. The only important thing is that the child is happy and is willing to let others participate in his or her singing!

Voice is a very sensitive instrument. Therefore, always stick to the most important rule: Never sing when your throat aches or is hoarse.

Welcome to Number Land!

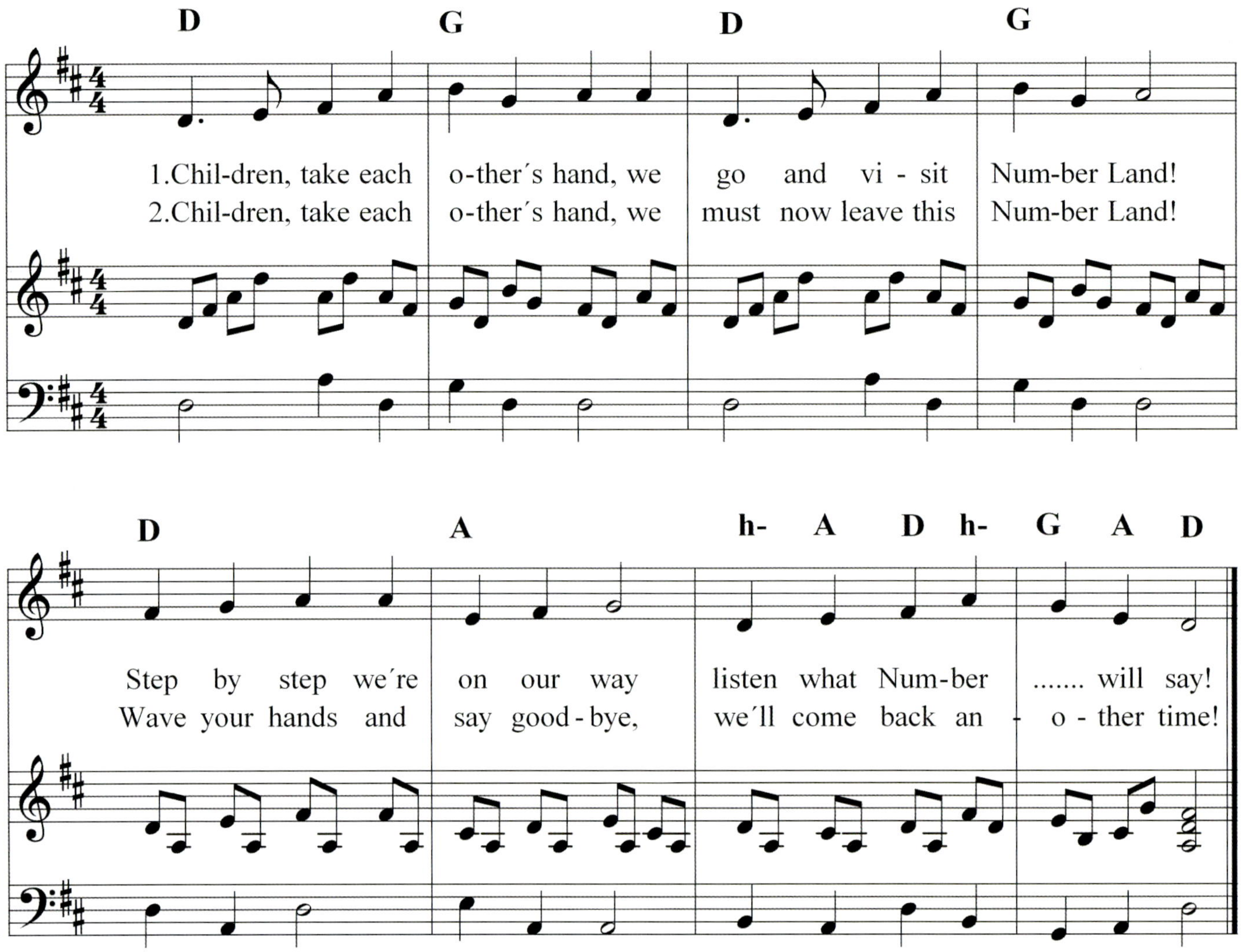

Lyrics and Music by Vio de Galgóczy

1st verse
This is the musical preparation for a walk in Number Land and the meeting with the next number. All children hold hands and walk, dance or jump/hop in a circle or are building a line which moves to the next number.

2nd verse (final refrain)
It leads back from fabulous Number Land into the merry and singing round. Children are waving theis hands in good-bye to the respective numbers, dance exuberantly or leave very quietly, in order not to wake up the dear numbers who just went to sleep.

Song of Number One

Music: 1/1-beat,
melody on one tone
Lyrics: one nose, first house

Lyrics and Music by Vio de Galgóczy

The melody of this song consists of one tone only except for a small closing. At first view this seems of small suspense, but rhythm and text are forming an unit of many harmonies leading like red thread from start to end.

songs: Teddy Bear / Sally Go ‘Round the Sun / The Mulberry Bush / Head, Shoulders, Knees and Toes / Twinkle, twinkle little star/ One Little Flower, One Little Bee

Song of Number Two

Music: 2/4-beat,
major and minor chords,
melody with two tones
Lyrics: dog+cat, sun+moon,
eyes+windows

Lyrics and Music by Vio de Galgóczy

This melody consists already of two tones, oriented like in the first song rhythmically on the text. The accompaniment changes blues-like from major to minor and back.

Songs: Two Little Dicky Birds / Pop! Goes the Weasel / Apples and Bananas / The Bear Went over the Mountain / Jack and Jill / Round and round the garden

Song of Number Three

Music: 3/4-beat, first 3 tones of the major scale
Lyrics: 3 steps, 3 tones, 3 times knocking

Lyrics and Music by Vio de Galgóczy

This song offers itself for a three-step-waltz, or a round-dance in groups of three, or to clapping hands at each turn of one.

Songs: Baa, Baa, Black Sheep
Action poems: Three Balls / Three Little Nickels

Song of Number Four

Music: 4/4-beat, four tones of the major scale
Lyrics: four seasons, four elements, four corners, four friends

Lyrics and Music by Vio de Galgóczy

This song is, because of its reminiscence of Reggae, very much preferred by children. A small whirligig can be made out of it by giving, before the singing begins, each child a special appointment like one of the four seasons or such. Each time a child is meant, it must stand up, turn around and around, hop or do some such a thing.

Song: Counting to Four (Leslie Feist singing in Sesame Street's 39th season)

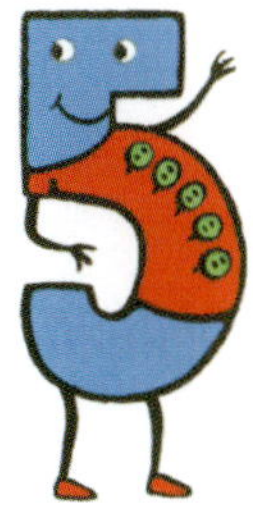

Song of Number Five

Music: 5/4-beat, five tones of the major scale
Lyrics: five corners / five fingers
Set theory: 5 are more than 3

Lyrics and Music by Vio de Galgóczy

Don´t be afraid, it is very easy to grasp this 2+3 quarter rhythm: The emphasis of the lyrics corresponds with the rhythm of the bass. In many countries, for example in Greece, these rhythms belong to everyday music life and are easily transformed by young and old into dance and song. Please clap your hands or stamp your feet according to the rhythm of the bass.

Songs: Five little Ducks / Five little Speckled Frogs / This Old Man / Hey, diddle diddle
Counting fingerplays or action poems: This little Pig / Where is Thumbkin? / Monkeys In A Tree Teasing Mr. Crocodile /Counting Apples / Five Fat Peas / Five Cookies / Five enormous dinosaurs

Song of Number Six

Music: 6/8-beat,
six tones of the major scale
Lyrics: six steps, six tones,
six rainbow - colors

Lyrics and Music by Vio de Galgóczy

This dreamy little waltz is not so easy to recite. Therefore it is advisable to start by describing the colors of a rainbow and painting one, reciting said colors fast and then dancing a colorful roundelay with colored ribbons.

Songs: All insects have 6 legs so any songs on lady bugs, butterflies, bees, ants and so on fit perfectly!
Example: The Ants Go Marching (Traditional action song from 1 to 10)

Song of Number Seven

Music: 7/4-beat, seven tones of the major scale
Lyrics: seven steps, seven chimes, seven flowers, seven days make one week

Lyrics and Music by Vio de Galgóczy

Once again: Don´t be afraid of the seemingly complicated rhythm! It can be split up into 4+3 quarters. The major 7, the seventh tone of the major scale, will be reached in the third and fourth bar. These bars are additionally emphasized by a small bell.

Songs: Head, shoulders, knees and toes (Seven body parts)

Song of Number Eight

Music: 8/8-beat, eight tones of the major scale
Lyrics: the numbers from 1 to 8

Lyrics and Music by Vio de Galgóczy

This funny little song presents the whole major scale, ending with an artistic jump from the 8th to the 1st tone. The 8th tone sounds higher than the 1st tone, in between lies one whole major scale.

Songs : Eensty Weensty Spider / Little Miss Muffet (action poem about a lady scared by a spider)

Song of Number Nine

Music: 9/8-beat
Lyrics: nine planets, nine children
nine steps, knocking nine times

Lyrics and Music by Vio de Galgóczy

The 9th tone can be transposed and can be changed into the 2nd tone of the major scale, so that it can be sung much easier. 9 eighths are corresponding to 3 quarters, which make a nice little waltz: nine children are dancing a roundelay.

Songs for all/ any numbers: 1, 2, 3, 4, 5, once I caught a fish alive (counting from 1 to 10) / This Old Man (counting from 1 to 5) / Ten little Indians / Monkeys On the Bed

Song of Number Ten

Music: 10/8-beat
divided into 2 x 5/8-beat
Lyrics: number 1 - 10,
ten steps, ten tones

Lyrics and Music by Vio de Galgóczy

Children may show right at the beginning of the last song of the Promenade of Numbers that they are mastering already the cardinal numbers: All numbers will be recited one after the other. Similar to song number 4, children may dance or hop around as they like best.

Songs: Ten green bottles (counting down from 10 to 0) / There were ten in the bed (counting down from 10 to 0) / One, Two, Buckle my Shoe (counting up to 10) / Ten little Indians (counting down from 10 to 0)